The Cotswold Sheep Flock Book with the Origin and History of Cotswold Sheep

Volume 1

by Cotswold Sheep Society

with an introduction by Jackson Chambers

Self Reliance Books

Get more historic titles on animal and stock breeding, gardening and old fashioned skills by visiting us at:

http://selfreliancebooks.blogspot.com/

Introduction

I am pleased to present yet another practical title on breeding and raising livestock.

The work is in the Public Domain and is re-printed here in accordance with Federal Laws.

As with all reprinted books of this age that are intended to perfectly reproduce the original edition, considerable pains and effort had to be undertaken to correct fading and sometimes outright damage to existing proofs of this title. At times, this task is quite monumental, requiring an almost total "rebuilding" of some pages from digital proofs of multiple copies. Despite this, imperfections still sometimes exist in the final proof and may detract from the visual appearance of the text.

I hope you enjoy reading this book as much as I enjoyed making it available to readers again.

Jackson Chambers

CONTENTS.

EDITING COMMITTEE.

———o———

Barton, Charles.

Brown, Thomas, *Chairman.*

Elwes, H. J.

Garne, Robert.

Hulbert, T. R.

Swanwick, Russell.

Yeomans, J. H.

THE
COTSWOLD SHEEP SOCIETY.

~~~~~~~~~~~~~~~~~~~~~~~

PRESIDENT:

ROBERT GARNE, Aldsworth, Northleach.

PRESIDENT-ELECT:

THOMAS BROWN, Marham Hall, Downham Market,
Norfolk.

COUNCIL:

CHARLES BARTON, Fyfield, Lechlade.

THOMAS BAGNALL, Westwell, Burford.

EDWARD BRAIN, Upper Slaughter,
Bourton-on-the-Water.

C. E. CLARKE, The Downs, Chalford.

H. J. ELWES, Colesborne Park, Cheltenham.

J. J. GODWIN, Troy Farm, Bicester.

T. R. HULBERT, North Cerney, Cirencester.

ROBERT JACOBS, Signett Hill, Burford.

WILLIAM LANE, Broadfield, Northleach.

THOMAS PORTER, Baunton, Cirencester.

RUSSELL SWANWICK, College Farm, Cirencester.

J. H. YEOMANS, Stretton Court, Hereford.

SECRETARY:

JAS. W. TAYLER, Cold Aston, Cheltenham.

PREFACE.

In presenting the First Volume of the Flock Book of the Cotswold Sheep Society, the Editing Committee congratulates the Members on the goodly number of Entries of Flocks and of Rams, but regrets that so few Members have entered Ewes— a defect in this Volume which will doubtless be remedied in succeeding Volumes. On the recommendation of the Editing Committee, the Council adopted the following Conditions of Entry for Vol. 1.

RAMS

Lambed before the 1st January, 1891.

That a Ram must be named and have been used in the Flock of the person entering the same.

EWES

Lambed before the 1st of January, 1890.

That a Ewe must be by a Registered Sire, form part of a Registered Flock, and have been bred from.

The Entry Fees charged to Members have been, for each Ram, 1/-; for each Ewe, 3d.; and for Registry of Flock, 5/-.

The Council empowered the Editing Committee to assign a Distinctive Letter to the Flock of each Member applying for the same. Twenty-two applications have been received and Flock Letters assigned accordingly, and the Editing Committee desire to point out that Distinctive Letters and Earmarks are absolutely essential to the Registration of individual Ewes.

The Entries in this Volume are—

Flocks... ... 45.

Rams ... ... 403 (numbered from 1 to 403).

Ewes ... ... 645.

In conclusion the Editing Committee believe that the Essay at the beginning of this Volume will be found to contain valuable information, and be appreciated by Members; and that by the publication of the Flock Book, now issued, the foundation has been laid for the future prosperity and improvement of Cotswold Sheep.

A PRIZE ESSAY

ON THE ORIGIN, HISTORY, AND PRESENT

POSITION OF THE COTSWOLD SHEEP.

COTSWOLD SHEEP.

Their Origin, History, and Present Position.

First Prize Essay by WILLIAM SCOTFORD HARMER,

Cirencester.

THE origin, history, and present position of Cotswold sheep is a three-fold subject which might well tax both the ability and the industry of the most experienced authority on the live stock of this country, but now that the producers of these famous animals have set seriously about the undertaking of giving their flocks their due place in agricultural annals and literature, the attempt may well be made to put on record some facts concerning a breed of sheep which, though one of the latest to have a Society and a Flock Book of its own, has no reason to fear comparison with any of its rivals in historical antiquity and in distinct characteristics.

ORIGIN AND EARLY HISTORY.

The origin of the Cotswold sheep is a problem on which much interesting matter has been published. Professor John Wilson, writing in the Journal of the Royal Agricultural Society in 1856, said that "the range of oolite hills running from North-east to South-west, and occupying the Eastern Division of Gloucestershire, have given their name to a breed of sheep which is probably one of the oldest recorded native breeds of the country." But the truth more probably is that, to quote Camden, the sheep give the name to the hills upon which they exist— "Cotes" meaning buildings in which they were formerly housed, and "Wold" being a wild open country, such as their habitation must in its primitive time have been. Therefore, with more literal accuracy than the phrase generally implies, Cotteswold or Cotswold sheep may be said to be "as old as the hills."

Nearly thirty years ago Mr. James Marsh Read, formerly of Elkstone, and himself a breeder of some repute, expended considerable research on the subject of the early history of the Cotswold sheep, and the facts which he elicited illustrate in a remarkable degree their historic interest as well as their intrinsic value. It is pretty clear that in the state of semi-barbarism of English life in the time of the ancient Britons, when the country abounded with wild animals, the typical domesticated animal, the sheep, had no place. In the words of Buffon, "the sheep was confided to the guardianship of man," and "needs his protection in order to subsist and his care to multiply." But soon after the Roman invasion had introduced civilization and the accompanying arts and luxuries of life to this island, there is ample evidence that sheep existed and were valued. Not only have sheep bones (which are absent from the burying places of ancient Britons) been discovered in Roman remains, but cloth was, without doubt, manufactured at Winchester and Gloucester by the Romans, while we learn from the writings of Tacitus that an extensive clothing trade existed at Cirencester early in the first century of the Christian era. Obviously this manufacture of cloth on the Cotswold hills involves the contemporaneous existence of sheep in the locality in order to supply the necessary raw material, and there is little room to doubt that the Cotswold sheep of the present day are the direct descendants of those which flourished on these hills eighteen or nineteen hundred years ago, and which were perhaps the type from which all our domesticated sheep have sprung. Taking it for granted that the sheep was not a domesticated animal of the ancient Britons, and that it was imported into England by the Phœnicians, or, as is more probable, by the Romans, the use of the the word "cote" as applied only to sheep on the Cotswolds is, as Mr. Marsh Read points out, very significant, implying as it does the Eastern derivation of the original Cotswold breed. The practice of "cotting" sheep is derived from the earliest times in the East, and the Biblical allusions to it, particularly in patriarchal times, are numerous. Saul, in seeking David, comes "to the sheep-cotes by the way where there was a cave."—1 *Sam.* xxiv. 3. "I took thee from the sheep-cote, from following the sheep."—2 *Sam.* vii. 8. Hezekiah had "stalls for all manner of beast and cotes for flocks."—2 *Chron.* xxxii. 28.

We now proceed to trace the history of the breed from early times to the present.

Although the scientific breeding of our domesticated animals, and the improvement of modern live stock by careful attention to the maxim that "like produces like," attained little prominence before the last century, the sheep of the Cotswolds attracted conspicuous notice at a very early date. Throughout Saxon and Norman times there is ample historical evidence that the keeping of sheep and the working up of the wool were carried on to a very large extent in the country around Cirencester, and an idea of the enormous flocks kept on these hills in the 13th century may be formed from the statement in Goding's "History of Beverstone" that "the quantity of sheep, nearly 6000, kept at Beverstone is remarkable," while numerous monumental brasses to woolmongers in Cirencester and Northleach churches attest the importance of the wool trade at the dates to which they refer. The immense quantity of wool cultivated in the county during the reign of Edward III. is apparent from the fact that 30,000 sacks of Cotswold wool was the annual quantity granted from the county of Gloucester for the King's household.—(Goding's "Ancient Trade of Wool and Cloth.") About the 14th century, the Florentines imported largely into this country, and in return for their goods "they carried away wool and cloth which they were accustomed to travel to Cotteswold to buy up" ("Pictorial History of England"). In the 15th century, it appears that both sheep and wool were largely exported, for in 1425 it was enacted by 3 Henry VI., c. 2, in order to remedy this state of things, "that no sheep shall be exported without the King's license," and there is no record of the King having been asked to grant a license permitting the exportation of the wool of any other than that of Cotswold sheep. In 1437 Don Duarte, King of Portugal, and brother-in-law of the King of Castile, from whom he might easily have obtained the choicest of Spanish wool, made application to Henry VI. for liberty to export sixty sacks of Cotswold wool, in order that he might manufacture certain cloths of gold at Florence for his own use. We learn from Stow that in 1464 a present of Cotswold rams was sent by Edward IV. to Henry of Castile ; and in 1468 twenty Cotswold ewes and four rams were shipped for John of Arragon. Markham, a writer on agricultural affairs in the time of Queen Elizabeth, says that the Cotswold sheep were, as they continued in every period of their early history, a long-woolled and large-boned breed. In Camden's "Brittania" we read

> "Where Cotswold hillocks famed for weighty sheep
> With golden fleeces clothed."

While Drayton, a little later wrote—

> "To tell
> How Evesham's fertile vale at first in liking fell
> With Cotswold, that great King of Shepherds,
> T'whom Sarum's plain gives place, though famous for her flocks,
> Yet hardly does she tythe our Cotswolds' wealthy locks."

Thus from the earliest times, when no other sheep were noticed, the Cotswold breed attracted conspicuous observation, and the evidence goes far to prove that they are the original type from which a considerable proportion of the Longwool sheep of the country have sprung; soil, climate, management, and the skill of the breeders having combined to produce the great changes in form and appearance displayed in other parts of the country.

IMPROVEMENT AND DIFFUSION OF THE BREED.

It is probable, both from recorded history and tradition, that the original unimproved Cotswold sheep was a large, flat-sided, somewhat leggy animal, with long heavy wool. But by a careful process of improvement and selection, the quality of the breed has been advanced without diminishing its size, and the Cotswold is now probably the largest, as he is certainly the hardiest, of our English breeds. The improvement to which we have referred, doubtless, to a great extent, took the form of a judicious infusion, during the latter part of the last and the beginning of the present centuries, of blood from the Leicester breed, for which Mr. Bakewell did so much about a hundred years ago. Among those to whom the country is indebted for the improvement of the breed may be mentioned the well-known names of Large, Garne, Hewer, Lane, Barton, Gillett, Walker, Fletcher, and many others.

Having always been most entirely in the hands of tenant farmers, who pursue farming as a business and not as a hobby, Cotswold Sheep have lacked the advantages which many other breeds have reaped from the support of wealthy patrons, who, regardless of expense, have brought their favourites before the world, and made them fashionable and popular. So for many years the merits of this handsome race were unappreciated, except upon their native hills. But when the desire for agricultural improvement manifested itself in the establishment of agricultural shows, the excellences of the breed attracted to it general and widespread attention, and the annual sales of the Cotswold ram breeders, which had for some years been established, were largely patronised. The result was that the sheep rapidly dispersed all over the British Isles, while a brisk foreign demand speedily sprung up, which has been well maintained to the present time. Amongst the counties which established pure-bred flocks (outside the hills of Gloucestershire and Oxfordshire) were Wilts, Hereford, Worcester, Glamorgan, Norfolk, Kent, Somerset, and others. One cause of the greatly extended use of these sheep was the property, which they possess in an eminent degree, of adapting themselves to almost every difference of climate and food. They flourish on their own poor exposed Cotswolds, and the richest pastures of Leicestershire and Buckinghamshire are not too much for their constitution.

The late Mr. John Bravender, over 40 years ago, writing on Cotswold farming, and the improved Cotswold sheep, with the production of which Cotswold farming is so largely concerned, says "in the time of Rudge, serious attempts were made to improve the breed of sheep, and which after discussion amongst breeders ended in producing the sheep so well-known to visitors at the annual meetings of the Royal Agricultural Society as the winners of all the prizes in the longwool classes. Our sheep are in great demand in all parts of England as well as Ireland for crossing with other breeds. Great pains have been taken to improve our sheep by many spirited breeders, and they have been well repaid for their trouble." Mr. R. Smith, in his prize essay on the management of sheep, wrote in 1847 much to the same effect, adding that Cotswold rams were "much sought after for crossing with short-woolled breeds, and with good effect."

An old breed of undoubted purity and great size, the merit of the Cotswold sheep for crossing with ewes of short-woolled and smaller boned varieties has long been very widely recognised, and there is hardly a breed in the country which has not at one time or other sought to improve its size and character, both in regard of mutton and wool, by the importation of Cotswold blood. Indeed, it should not be forgotten that the Cotswold sire is the parent of one of the best, as it is certainly the largest, of our Down breeds. The Oxford Downs, of comparatively modern origin, were introduced from a cross of Cotswold rams with Hampshire Down ewes, and they owe their large and handsome frames to their Cotswold progenitors. Although, in 1857, it was decided at a meeting of flockmasters held at Oxford, to give the breed thus formed the name of Oxfordshire Down, and the Royal Agricultural Society assigned a distinct class to it under that name in 1862, Mr. Clare Sewell Read, in his prize report on the farming of Oxfordshire, gave it the name of Down-Cotswold, a nomenclature which prevailed for some time, and which at any rate possessed the merit of recognising the source from which the early maturity, heavy carcase, and ample fleece of the new breed were derived. The change of the name from Down Cotswold to Oxfordshire Down was not allowed to pass without a protest from the Cotswold men, and this was, perhaps, but natural, as but a few weeks before the new name was formally adopted a number of rams had been sold in Oxford market as "Oxford Downs" which had been got by a Cotswold ram.

With their animal much improved and gaining ground in public favour by the force of its own merits, the number of breeders of Cotswolds greatly increased, till some forty years ago it was estimated that in Gloucestershire alone 5000 rams were sold and let in a season at a total price of little less than £50,000. Some ten years later the number of rams sold annually was put at about 4,000, and at that time there was a good export trade to America, Australia, and the continent of Europe, as well as a brisk demand from all parts of the United Kingdom. The numbers now bred are not so large, but the names of many famous breeders of that time still survive.

COTSWOLDS AND THE ROYAL AGRICULTURAL SOCIETY.

A few words as to the position which Cotswold sheep have taken in the national show yard may not be out of place. For the first ten or twelve years of the existence of the Royal Agricultural Society of England, the only classes for sheep at their annual shows were (with now and then an exception for local reasons) those for Leicesters, Southdowns, and Longwools not Leicesters. Cotswold men of course showed in the last-named class and invariably swept the board of the prizes, and in the earlier days of the Society the names of Large, Garne, Lane, E. Smith, Hewer, Handy, Beman, Wells, Slatter, Fletcher, Beale Browne, Walker, King Tombs, Gillett, and others were amongst the most prominent prize winners.

In 1851, when the great show of the Royal Agricultural Society was held at Windsor, the Council considerably restricted the prizes offered for Cotswold sheep, the premiums being much below those given for other breeds. A remonstrance from the Gloucestershire breeders was prepared by Mr. W. Cother, of Middle Aston (member of the firm of Lyne and Cother, the well-known Cotswold sheep auctioneers), and forwarded to the Society, pointing out the large and increasing extent to which Cotswold sires were imported into almost every county in England to improve the size of other sheep. Mr. Cother asked what valid reason there was for cutting down the prize list in their class, and added : "I have reason to know that Cotswold sheep have gone from our hands into all countries where any other English sheep have gone, and a goodly number of breeders who once resorted to the flocks of the honoured names of Stubbins, Burgess, Burdon, Farrow, Stone, &c., are now constant customers to us." Probably "the honoured name of Stubbins" was too much for the Council, for at the meeting in 1853, held at Gloucester, the class for "Longwools not qualified to compete as Leicesters" was restored to its old position in the prize list. In 1854, commenting on the sheep classes at the Royal Show that year held at Lincoln, *The Times* said : "At one period the large-framed, well-fleeced, and hardy Cotswold breed were systematically branded by the Society as inferior stock, from the smaller prizes awarded to them. But a vigorous resistance on the part of the breeders, and a threat that they would take steps to introduce fresh blood into the Council, induced that body to do them justice. The consequence has been that the Cotswold sheep is introduced to the notice of the agriculturist in this part of the country with the same prestige as Leicesters and Southdowns. He is enabled to compare it with the 'Improved Lincoln,' and judging by the results of the show yard it possesses a manifest superiority. And further, he has an opportunity of considering how far an animal which thrives so well on the uplands of Gloucestershire would answer on the heaths and wolds of Lincolnshire. No less than 19 Cotswold farmers exhibited their stock on this occasion. Such enterprise is highly commendable, and has probably been stimulated by the increasing demand for the breed from abroad. It has found great favour with American agriculturists especially, and I am informed that one flockmaster has this year sold seven rams and ten ewes to Transatlantic purchasers at prices in the aggregate amounting to nearly £1000."

About the year 1867, the older Cotswold breeders, who had hitherto gained the chief honours at the Royal shows, discontinued exhibiting. The result was that the breed was not so well represented in the show-yard as formerly was the case. In fact, as the record of sales shows, Cotswolds had now obtained such a well-established position that they could afford to dispense with the assistance which the " Royal" had undoubtedly rendered them in the earlier years in order to bring their merits before the public. Writing on this subject in 1869, the *Agricultural Gazette* said :—" The Cotswold breed is now rarely represented according to its merits. The noblest breed in the country is in the hands of a somewhat quiet and homely body of agriculturists—perhaps more averse than most to the quackery and puffery of mere display." But though the older breeders retired, there was not wanting young blood to take their places, and of late years such exhibitors as Mr. T. Brown, Mr. Russell Swanwick, Mr. G. Bagnall, Mr. T. R. Hulbert, and others have furnished the Royal show-yard with worthy specimens of the breed.

AN IDEAL COTSWOLD.

Leaving now the realm of general history, and coming to matters of commercial interest, it may be convenient to state here the points of an ideal Cotswold sheep. To start with that important feature, the head— a point on which Mr. Henry Corbet laid such stress as an index to breed and quality, and about which he gossiped so pleasantly twenty years ago in the Journal of the Bath and West of England Society ; the head should be wide between the eyes, and the eye itself, full, dark, and prominent, but mild and kindly, and in no way coarse about the brow. The face should be proportionately wide to the space between the eyes, but not too flat, and should run of much the same width to the nostrils, which must be well expanded and somewhat broader than the face, with the skin on the nose of a dark colour. The cheek is full, and, as is the face, well covered with white hair; a just perceptible blue tinge on the cheek and round the eye being rather "fancied." The ear, long, but not heavy, of medium thickness, and covered with the same short soft hair; should be well carried up, while black spots on the point of the ear are not considered objectionable. The top of the head should not be coarse nor bald, but covered with *wool*, not hair, and the Cotswold is to be distinguished by a fine tuft of wool on the forehead. The head should be sufficiently long to save it from being called short and thick, but it should not have a long lean appearance. Grey faces still crop up occasionally in all the best Hill flocks. The neck should be big and muscular, and should be gently curved to enable the sheep to carry the head well up, thereby giving the animal a grand appearance. The neck should be slightly smaller at the ears than where it comes from the shoulders. The shoulders should lay well back, and the point of the shoulder should be well covered with flesh, as also the chives. The ribs should be deep, well sprung from the back; the hips and loin wide and well covered with flesh. The rump should be carried out on a level with the back, giving the animal a square-looking frame; the leg of mutton well let down to the hock, and thick on the outside. The legs, both front and hind, should be straight, moderate in length, well set outside the body. The pastern joints, both front and hind, should be short. The whole body should have a firm, solid touch (not loose and flabby), and be well covered with a thick-set, long, and lustrous wool.

MUTTON AND WOOL PRODUCTION.

Cotswold sheep are capable of enduring great hardships, succeeding well in exposed situations, and on nearly all kinds of soil adapted for sheep farming. They produce a great amount of mutton and wool at an early age, their rapid maturity and disposition to fatten enabling them to be brought to market, at from 9 to 12 months old, with ordinary feeding, at a weight of from 90 to 112 lbs., while it is no unusual thing for the best flocks to turn out 120 to 130 lb. sheep at that age. Indeed, at Christmas, 1853, as the result of a sweepstakes in which several breeders joined, Mr. T. T. Porter, of Baunton, produced a pen of ten Cotswold tegs whose ages were between 8 and 9 months only, and whose average weight was 30 lbs. per quarter, the heaviest weighing 139 lbs. In this connection the following newspaper extract *(Wilts and Gloucestershire Standard, 1851)* may be of some interest: "Ten Cotswold wether sheep under 14 months old were slaughtered at Chipping Norton on the 30th April; they were weighed in the presence of scores of spectators, and averaged rather over 30 lbs. per quarter (the net four quarters), sinking a little less than one-third their live weight, and clipping 13 lbs. of clean washed wool each." The breeder and feeder of these sheep was Mr. Coldicott, of Over Norton, near Chipping Norton, the sires being bred by Mr. John Gillett, of Minster Lovell. As showing the enormous weight to which Cotswolds can be brought as old sheep, in the following year one bred and fed by Mr. Cother, of Middle Aston, was killed at the age of 3 years and 9 months, and weighed 336 lbs., or 84 lbs. a quarter, one of the legs of mutton weighing 54 lbs. At Christmas, 1859, Mr. Robert Garne produced a sheep weighing 43 stones, or 86 lbs. per quarter, for which he obtained £8 10s. In 1856, the *Yorkshire Gazette* chronicled the exhibition at Market Leighton market of Mr. T. Beale Browne's three-years-old Cotswold ram, "Champion," which took prizes at the leading shows in England and Ireland as well as in Paris, and said he was "considered by judges to be the most wonderful animal of the sheep kind ever seen in Yorkshire. His estimated weight is about 100 lbs. per quarter." Such instances of massive proportions could be largely multiplied, but these are given by way of sample records. The meat, especially when young, is succulent and well flavoured.

Of course, with such a pronounced Longwool breed, the wool produce is an important item in a Cotswold flock, and it may be safely asserted that the Cotswold sheep grows a fleece of a special character which, for weight and substance, no other part of the world can rival. The staple of the wool is long and mellow, and the average weight of the fleece throughout a well-managed first-rate flock would be from 9 lbs. to 11 lbs. of washed wool.

A chief feature in the valuable characteristics of the Cotswold sheep is his handsome return for the food expended on him. This point is so well established that it is hardly necessary to dwell upon it, but it may be mentioned that so long ago as 1856 Sir J. B. Lawes (Mr. Lawes as he then was) published in the Journal of the Royal Agricultural Society the results of exhaustive and elaborate "experiments on the comparative fattening qualities of different breeds of sheep." The sheep experimented upon were Hampshire Downs, Sussex Downs, Cotswolds, Leicesters, Cross-bred wethers, and Cross-bred ewes. The results showed that "the Cotswolds consumed the least food to produce a given amount of increase." In summing up his results Mr. Lawes said: "The Longwools, especially the Cotswolds, will yield a larger amount of gross increase for a given amount of food than the Downs or Crosses. The average prices of Down and also of Cross-bred mutton and wool are higher than those of the Longwools, but not sufficiently so to compensate for the cost of the extra food consumed. It would appear, therefore, that when equally fitted to climate, locality, and system of farming adopted, and when what may be termed a fancy price for Down mutton is not attainable, those animals yielding the most mutton and wool from a given quantity of food will have an advantage in supplying the demand of the masses of the population." Of course, in these days of importation of sheep and mutton from the other ends of the world, the observation as to "supplying the demand of the masses of the population" loses much of its significance, but that the Cotswold still remains the cheapest mutton-producing machine that can be located on his native hills is beyond all question, for subsequent feeding experiments have repeatedly confirmed Sir J. B. Lawes's conclusions. In a trial made by Lord Kinnaird, Cotswolds fed against Leicesters showed, with the same quantity of food, a gain in value of 17s. against 11s. 8¼d. gained by the Leicesters. This result confirmed one of the earliest feeding experiments we have come across, in which the same two breeds were pitted against each other. In an old account of this experiment we read : " In consequence of a challenge from Mr. Moore, of Charlcot, Warwickshire, to Mr. Peacey, of Northleach, at the meeting of the Bath and West of England Society in December, 1791, a wager took place between those gentlemen, on the following terms : Mr. Moore agreed to send five shearhogs of the Leicestershire, and Mr. Peacey five of the same age of the Cotswold, with a small cross of the Leicestershire in them, to John Billingsley, Esq., of Ashton Grove. They were to be weighed when settled to their pasture, kept together exactly alike; and again weighed previous to the December meeting in 1792, and then slaughtered and the carcases exhibited for the inspection of the Society. The sheep that gained most weight during the time of trial, and whose wool was the most valuable were to be deemed the best and win the wager. The following are the particulars of their weights, &o.

	Mr. Peacey's.	Mr. Moore's.
Jan. 3, 1792	697½ lbs.	664½ lbs.
Dec. 8, „	952½ „ :... ...	762 „
Gained in weight	255 lbs.	97½ lbs.

Superior increased weight of Mr. Peacey's 5 sheep, 175½ lbs.

	£	s.	d.
Mr. Peacey's wool, 36¼ lbs. at 12d.	1	16	6
Mr. Moore's „ 19⅜ „ „ 9d.		14	6
In favour of Mr. Peacey's	£1	2	0

COTSWOLD RAM SALES.

The records of the sales of Cotswold sheep are voluminous, and can only be treated here in outline. The general favour in which they were received some half century ago, and which they have maintained, with some fluctuations, ever since, has enabled the foremost breeders to obtain handsome prices for their stock. Whether Justice Shallow, who without doubt, as a Cotswold squire, kept his Cotswold flock, held an annual ram sale, Shakespeare does not tell us. If he did, his averages were probably very different to those obtained by his 19th century successors. A notion of the value of Cotswold sheep in the early part of the 14th century may be gained from a glance at Squire Shallow's conversation with his cousin Silence at the former's Gloucestershire seat :—

Shallow : How a score of ewes now ?

Silence : Thereafter as they be; a score of good ewes may be worth ten pounds.

The values of both sheep and money are very different in the days of Queen Victoria to what they were in the days of Henry IV., and we will come down to more recent times. As to the origin of the Cotswold ram sales, we believe that the Mr. Peacey who held for many years Hill House Farm, Northleach, which Mr. W. Hewer afterwards held and rendered famous, and who has been before referred to as pitting his sheep against the Leicesters for feeding qualities, was the first to start these sales, and that he did so as early as about 1785. The sales were at first conducted by private contract, breeders announcing the dates on which their rams would be ready for inspection, and their customers then assembling and dealing for the sheep which met their fancy. Later on the practice of holding an auction at the homestead was resorted to as a more expeditious mode of disposing of the business, while at the same time introducing a healthy spirit of competition between those who had set their heart on any particular lots. In years gone by, the series of ram sales on the hills formed an extensive round of social celebrations and hospitalities which occasioned large and festive reunions, but as time went on, and the auction sales at fairs and markets increased in public favour, general resort was gradually had to the large Sheep and Ram Fairs held at Cirencester in August and September, as well as, though to a much lesser extent, to similar fixtures at Oxford, Gloucester, and other towns. The sales at the homesteads were thus one by one discontinued, till the Aldsworth sale (Mr. Robert Garne's) is now the only one held at home on the Cotswold Hills, though in Norfolk Messrs. Brown & Son and Mr. Hugh Aylmer also still adhere to the old custom.

The honour of gaining the highest recorded average belongs to the late Mr. Robert Lane, of The Cottage Farm, Northleach, who in 1861 made £34 10s. 8d. a head. In 1867, Mr. W. Lane obtained an average of £31 17s. 11d., and in 1873, Mr. R. Garne realised £28 16s. 4d.

Mr. W. Hewer, of Hill House Farm, Northleach, was a very successful breeder, and gained great distinction both in the show and sale rings. One of his rams was sold in 1864, at 230 guineas, to Mr. W. Lane, of Broadfield, after a spirited competition with Mr. Charles Barton, of Fyfield.

TYPICAL FLOCKS.

Though Cotswold sheep are now for the most part confined to their native Gloucestershire and Oxfordshire hills, they have for many years been bred with much success in Norfolk and Hereford, and some pure flocks are also to be found in South Wales, and perhaps a few notes on one or two of the best known flocks at present in existence may be of interest. In view of objections in certain quarters to the possibility of keeping a Flock Book for any breed of sheep which shall be even approximately accurate, one general remark may be permitted, viz., that such a criticism applies with much less force to Cotswold sheep than to any other breed, for the leading Cotswold flockmasters conduct their breeding operations with scrupulous care, and they have for many years kept most exact records and registers, having in fact been in the habit of supplying well authenticated pedigrees of their sheep to such of their customers as required them.

Well, in referring to a few typical flocks, we naturally begin with the celebrated

ALDSWORTH FLOCK, in the possession of the first President of the Cotswold Sheep Society, Mr. Robert Garne. This flock was of considerable note in the latter part of the last century, when it was owned by the grandfather of the present proprietor, Mr. William Garne, who then lived at the neighbouring village of Sherborne, in partnership with a friend; this partnership was dissolved in 1800, Mr. W. Garne retaining his portion of the flock, and the other portion being sold in the spring of that year, making £5 a couple, to Mr. Joseph Large, who subsequently, and also his son, Mr. Charles Large, became celebrated breeders at Broadwell. Later on, say the latter part of the first half of this century, Mr. Large bid Mr. W. Garne, £6 6s. a couple for the portion of the flock he retained in 1800, which offer was refused. In the year 1801, the flock passed into the hands of Mr. R. Garne's father, who owned it until 1857, since which date it has been in Mr. R. Garne's possession. There is no doubt an improvement was effected by a cross with the Leicester breed, for the first Mr. W. Garne in the last century, and his son and successor in the early portion of this century, occasionally used Leicester rams, while the Messrs. Large referred to above infused a considerable dash of Leicester blood into their flock, and most of the Cotswold breeders used sheep from Messrs. Large. The Aldsworth flock has never exhibited at any show but the "Royal," whose show yard it entered at York in 1848. From that time to 1863 the sheep were sent yearly, and gained 25 prizes in the 13 years for rams, as well as two first prizes and a medal for ewes, having exhibited in this latter class four or five times only. In 1863, at Worcester, Mr. Garne won the whole of the prizes in the yearling class, in a competition of over fifty. Since then showing has been discontinued, with one exception, that being the occasion when the great show was held in Windsor Park, in 1889. Mr. Garne's father, Mr. William Garne, began selling by auction, in 1844, and the Aldsworth sale has been continued ever since without intermission. The highest average was in 1873, when 54 sheep made £28 16s. 4d. each.

THE BROADFIELD FLOCK, Mr. W. Lane's, and which is still carried on by his son-in-law and himself, has a reputation second to none on the hills. Mr. Lane founded his flock between forty and fifty years ago, commencing with some good sires from Messrs. Large Broadwell), W. Garne (Aldsworth), and W. Hewer (Northleach). He afterwards made a fortunate purchase from Mr. Eeles, of the Cottage Farm, of a ram called "The Broken Jawed Sheep," hired one and bought another of Mr. Hewer, and bought one from the late Mr. John Barton, of Fyfield. All these made a great mark, as did another good one from Mr. Eeles, called "Bandy." Mr. Lane followed on with the best sires he could get from all the famous flocks, including those of Messrs. Large, Garne, Barton, Fletcher, Handy, Gillett and others, giving from 70 guineas and 90 guineas, up to 230 guineas, buying a ram at the last-named figure from Mr. W. Hewer, of Northleach, which paid him very well. Mr. Lane at first held private sales, but embarked on an annual auction in 1852, which he continued without interruption for close on 40 years. His sales usually included from 50 to 60 shearlings, and the highest average he ever gained was £31 17s. 11d. for 54 sheep in 1867. In 1866, when Mr. W. Lane's 54 sheep averaged £26 18s. 9d., four of the rams made upwards of £100, one going to Mr. J. King Tombs at 210 guineas, another to Mr. G. Fletcher at 122 guineas, a third to Mr. T. Porter at 126 guineas, and a fourth to Mr. R. Garne at 100 guineas. Mr. Lane began showing at the meetings of the Royal Agricultural Society in the same year as he embarked on public auctions, viz., in 1852, and he continued to be represented for nearly 20 years, never coming back without several prizes. For many years he was one of the most successful exhibitors, not only in the ram but also in the ewe classes. Like many of his neighbours, Mr. Lane had many Transatlantic customers, and on one occasion he sold three lambs at £70 each, and another at 60 guineas. One of his patrons, Colonel Ware, was ruined at the time of the American war, and was utterly unable to pay for a considerable number of valuable sheep he had purchased. Mr. Lane with characteristic generosity, anxious to do what he could to set his old friend on his legs again, offered to make him a present of a ram and three ewes to make another start, but the Colonel had not the heart to make the attempt. He wrote saying that he had lost everything except one solitary turkey, and could not see his way to making another venture.

FYFIELD FLOCK.—Mr. Charles Barton's flock at Fyfield is one of the oldest, if indeed it is not the oldest on the Cotswolds. It was removed to Fyfield from Coln Rogers in 1828, by Mr. Barton's father, Mr. John Barton, and there are records extant of its having been in the possession of the Barton family at Coln Rogers from the early part of the 17th century, certainly from 1640 onwards. Like those of most of the old breeders, the Coln Rogers and Fyfield ram sales were for many years conducted privately, and it was not till the death of Mr. John Barton that in 1854 sales by auction were resorted to, and have since been continued. The sales at Coln Rogers began about 1819. Mr. Charles Barton, following the the example of his father, has rigidly abstained from entering the show-yard. The highest Fyfield average was £16 13s. 4d. in 1862.

ROYAL AGRICULTURAL COLLEGE FARM FLOCK.—Mr. Russell Swanwick's flock of Cotswold sheep on the Royal Agricultural College Farm was commenced in 1869 and 1870 by the purchase of 115 of the best ewes and theaves at the dispersion of Mr. William Hewer's famous Cotswold flock. Then on the death of that famous breeder of Cotswolds, Mr. Robert Lane, of The Cottage Farm, Mr. Swanwick purchased several pens, the cream of the flock, at almost unheard of prices, one pen at 17 guineas a head. To these were added theaves from Mr. William Lane's celebrated Broadfield flock. Mr. Swanwick now determined to use the best rams obtainable, and purchased from Mr. W. Lane, at his sale in 1870, a ram at 80 guineas, since known as "The Eighty Guinea Ram;" this ram was by Mr. Hewer's famous 230 guinea ram, and out of a ewe whose produce had up to that time averaged over 80 guineas each, she lived till she was over 12 years of age, showing the strong vitality of this family: this ram, after being used for 4 years, had increased the average weight of fleece per head of the whole flock by 2 lbs.; then followed numerous other purchases of rams from Mr. W. Lane, Mr. Fletcher, Mr. Robert Garne, Mr. Charles Barton, Mr. R. Jacobs, and Mr. Hugh Aylmer, showing the flock to contain the best of Cotswold blood. In selecting rams great care has been taken to keep the sheep of firm deep flesh, with good backs and good legs of mutton, and at the same time to keep the fleece of high quality and lustre, which will account for the flock having won numerous prizes for fleeces of wool. For the last twenty years, with the exception of one year, sheep have been exhibited every year, with the result that Mr. Swanwick has won £2,150 in prizes, by sheep bred by himself.

THE MARHAM HALL FLOCK.—The Norfolk Cotswolds have for close on 30 years been celebrated for their size and quality. The Marham Hall Cotswold Flock (founded by Mr. Thomas Brown, president-elect of the Cotswold Sheep Society, and now carried on by that gentleman and his son), had its origin from ewes bought in the years 1863-1871 of the late Mr. Robert Lane, the Cottage Farm, Northleach, whose flock was one of the oldest, most noted, and most successful on the Cotswold Hills. A few ewes have also been bought of the late Mr. William Hewer, Northleach, Mr. R. Garne, Aldsworth, Mr. William Lane, Broadfield, and others. The rams used have either been bred on the farm, or obtained from the sales on the Cotswold Hills, almost wholly from those of Messrs. R. Lane, R. Garne, William Lane, and William Hewer. Twice has a ram been bought at 120 guineas, and twice at 101 guineas. Specimens from this flock were first exhibited at the Royal Agricultural Society's shows in 1867, the meeting of that year being held at Bury St. Edmunds, and gained at that show 1st and 2nd prizes for Shearling rams, 1st and 2nd for older rams, and 1st for Shearling ewes; and from that time continuously until 1883, when systematic exhibition was discontinued, obtained the lion's share of the prizes, four times during that period carrying off all the prizes for Shearling rams, namely, at Wolverhampton, Hull, Bedford, and York. For many years upwards of 300 rams and ram lambs have been *let* and sold annually for breeding purposes, a letting having been held in every year late in July, or early in August, at which the Ram lambs have usually averaged from 6 to 8 guineas, and the Shearling rams from 10 to 12 guineas. On one occasion, 1883, 80 Ram lambs were *let* at an average of £9 6s. 9d., and 80 Shearling rams were let at an average of £15 4s. 3d. About 60 Shearling rams have been yearly sold by auction at Hempton fair, held on the first Wednesday in September, and have averaged from 10 to 15 guineas each. Upwards of a hundred lambs and shearlings have also been *let* or sold by private contract. A system of numbering by ear marks was adopted in 1883, and every lamb bred in that and subsequent years has been so numbered, and entered in a private flock book. Previous to 1883 only the pedigree of the rams used in the flock had been registered.

Several other flocks are equally deserving of notice, but these will perhaps suffice as giving a general idea of the History of the Cotswold Flocks of the present day.

THE
COTSWOLD SHEEP SOCIETY
FLOCK BOOK.

FLOCKS OF
Registered Cotswold Sheep,

TO WHICH

DISTINCTIVE LETTERS HAVE BEEN ASSIGNED.

---+---

<table>
<thead>
<tr><th>Flock Letter.</th><th></th><th>Owner.</th><th></th><th></th><th></th><th>Page.</th></tr>
</thead>
<tbody>
<tr><td>A. ...</td><td>...</td><td>... ROBERT GARNE</td><td>...</td><td>...</td><td>...</td><td>... 11</td></tr>
<tr><td>B. ...</td><td>...</td><td>... MAJOR THE HON. L. BYNG</td><td></td><td>...</td><td>...</td><td>6</td></tr>
<tr><td>B. F.</td><td>...</td><td>... CHARLES BARTON</td><td>...</td><td>...</td><td>...</td><td>... 4</td></tr>
<tr><td>C. ...</td><td>...</td><td>... ERNEST CRADDOCK</td><td>...</td><td>...</td><td>...</td><td>... 7</td></tr>
<tr><td>E. ...</td><td>...</td><td>... H. J. ELWES</td><td>...</td><td>...</td><td>...</td><td>... 9</td></tr>
<tr><td>E. E.</td><td>...</td><td>... EARL OF ELDON</td><td>...</td><td>...</td><td>...</td><td>... 8</td></tr>
<tr><td>F. ...</td><td></td><td>... THOMAS WALKER</td><td>...</td><td>...</td><td></td><td>17</td></tr>
<tr><td>G. F.</td><td>...</td><td>... GEORGE FREEMAN</td><td>...</td><td>..</td><td>...</td><td>10</td></tr>
<tr><td>H. ...</td><td>...</td><td>.. T. R. HULBERT ...</td><td>...</td><td>...</td><td>...</td><td>14</td></tr>
<tr><td>H. A.</td><td>...</td><td>... HUGH AYLMER ...</td><td>...</td><td>..</td><td>...</td><td>... 4</td></tr>
<tr><td>H. B.</td><td>...</td><td>... WILLIAM HOULTON</td><td>...</td><td>...</td><td>...</td><td>... 13</td></tr>
<tr><td>H. H.</td><td>...</td><td>... H. T. HOULTON ...</td><td>...</td><td>...</td><td>...</td><td>... 13</td></tr>
<tr><td>I. ...</td><td>...</td><td>... J. P. WAKEFIELD</td><td>...</td><td>...</td><td>...</td><td>... 17</td></tr>
<tr><td>J. ...</td><td>...</td><td>... ROBERT JACOBS</td><td>...</td><td>...</td><td>...</td><td>... 14</td></tr>
<tr><td>K. ...</td><td>...</td><td>... MICHAEL BIDDULPH ...</td><td>...</td><td>...</td><td>...</td><td>5</td></tr>
<tr><td>M. ...</td><td>...</td><td>... THOMAS BROWN & SON</td><td>...</td><td>...</td><td>...</td><td>6</td></tr>
<tr><td>P. ...</td><td>...</td><td>... WALTER POWELL</td><td>...</td><td>...</td><td>...</td><td>15</td></tr>
<tr><td>R. ...</td><td>...</td><td>... JAMES TAYLOR ...</td><td>...</td><td>...</td><td>...</td><td>16</td></tr>
<tr><td>R. S.</td><td>...</td><td>... RUSSELL SWANWICK ..</td><td>...</td><td>...</td><td>...</td><td>15</td></tr>
<tr><td>W. ...</td><td>...</td><td>... GEORGE BAGNALL & SON</td><td>...</td><td>...</td><td>...</td><td>4</td></tr>
<tr><td>Y. ...</td><td>..</td><td>... J. H. YEOMANS ...</td><td>...</td><td></td><td>..</td><td>18</td></tr>
<tr><td>Z. ...</td><td>...</td><td>... FREDERICK CRADDOCK</td><td>...</td><td>...</td><td>...</td><td>8</td></tr>
</tbody>
</table>

FLOCKS OF

REGISTERED COTSWOLD SHEEP,

ARRANGED ALPHABETICALLY

ACCORDING TO THE OWNER'S NAME.

ACOCK, ARTHUR,

Cold Aston, Cheltenham, Gloucestershire.

27 Yearling Ewes.

23 Two-year-old Ewes.

39 Older Ewes.

This Flock was established in 1861 by Ewes bought of — Powell, Cold Aston.

ATTWATER, J. N.,

Sireford, Andoversford, Gloucestershire.

71 Yearling Ewes.

61 Two-year-old Ewes.

108 Older Ewes.

This Flock was established in 1872 by the purchase of Ewes from G. Fletcher, Shipton; E. Handly, Sireford; and T. Walker, Compton Abdale.

AYLMER, HUGH,

West Dereham Abbey, Stoke Ferry, Norfolk.

100 Yearling Ewes.

100 Two-year-old Ewes.

200 Older Ewes.

Flock Letter, H. A.

BAGNALL, GEORGE, & SON,

Westwell Manor, Burford, Oxon.

130 Yearling Ewes.

135 Two-year-old Ewes.

235 Older Ewes.

Flock Letter, W.

This Flock was established by present owner's forefathers, and has been in the family without intermission for the last five or six generations.

BARTON, CHARLES,

Fyfield, Lechlade, Gloucestershire.

131 Yearling Ewes.

96 Two-year-old Ewes.

174 Older Ewes.

Flock Letter, B. F.

This Flock was brought from Coln Rogers, Gloucestershire, by present owner's father, John Barton, in 1828, and previous to that date had been maintained at Coln Rogers by his father, grandfather, and earlier generations of the Barton family for upwards of 200 years.

5

BEAK, GEORGE,
Stanford Hall, Lechlade.

120 Yearling Ewes.

100 Two-year-old Ewes.

120 Older Ewes.

This Flock was established by the present owner's father in 1804.

BIDDULPH, MICHAEL,
Kemble, Cirencester.

100 Yearling Ewes.

100 Two-year-old Ewes.

150 Older Ewes.

Flock Letter, K.

This Flock was established by Ewes purchased about 1889 from the Flocks of J. M. White, Kemble; Hulbert, North Cerney; and Craddock, Ablington.

BRAIN, EDWARD,
Manor Farm, Upper Slaughter, Bourton-on-the-Water.

50 Yearling Ewes.

50 Two-year-old Ewes.

73 Older Ewes.

BROWN, THOMAS, & SON,
Marham Hall, Downham Market, Norfolk.

121 Yearling Ewes.
91 Two-year-old Ewes.
227 Older Ewes.

Flock Letter, M.

This Flock was established principally by Ewes bought of Robert Lane, The Cottage Farm, Northleach, during the years 1863 to 1871, both inclusive, but a few were bought of R. Garne, W. Hewer, J. H. Langston, G. Garne, J. H. Pedley, and W. Lane.

BYNG, MAJOR HON. L.,
Sherborne House, Northleach, Gloucestershire.

60 Yearling Ewes.
70 Two-year-old Ewes.
72 Older Ewes.

Flock Letter, B.

This Flock was established by Ewes bought about 1889 of J. Wakefield, Barrington; E. Brain, Upper Slaughter; and C. Barton, Coln Deans.

CLARK, CHAS. E.,
The Downs, Chalford, Stroud, Gloucestershire.

63 Yearling Ewes.
70 Two-year-old Ewes.
86 Older Ewes.

This Flock was established nearly a century ago, by — Wymore, and came into the possession of the present owner in 1888, and has been replenished by Ewes purchased from C. Barton, Coln Rogers, and others.

CLARK, HENRY,

Frampton Mansell, Stroud.

150 Yearling Ewes.
110 Two-year-old Ewes.
120 Older Ewes.

This Flock was established in 1852.

COOK, THOS.,

Taddington, Broadway, Worcestershire.

71 Yearling Ewes.
60 Two-year-old Ewes.
113 Older Ewes.

This Flock was established about 1800, and came into the possession of the present owner in 1873.

CRADDOCK, ERNEST,

Macaroni Farm, Eastleach, Lechlade.

50 Yearling Ewes.
46 Two-year-old Ewes.
129 Older Ewes.

Flock Letter, C.

This Flock was established about 1867 by the father of the present owner, and came into the possession of the latter in 1889.

CRADDOCK, FREDK.,
Eastington, Northleach, Gloucestershire.

80 Yearling Ewes.
90 Two-year-old Ewes.
150 Older Ewes.

Flock Letter, Z.

This Flock has been in existence on the same Farm since 1790, and was probably established many years previously.

' ELDON, EARL OF,
Manor Farm, Chedworth, Northleach.

130 Yearling Ewes.
116 Two-year-old Ewes.
232 Older Ewes.

Flock Letter, E. E.

This Flock was established in 1882 by Ewes bought of — Sutton, — Radcliffe, and R. Minchin.

ELDON, EARL OF,
Compton Abdale, R. S. O., Gloucestershire.

178 Yearling Ewes.
151 Two- year-old Ewes.
217 Older Ewes.

Flock Letter, E. E.

This Flock was established in 1881, by Ewes bought chiefly at Joseph Walker's sale.

ELWES, HENRY JOHN.

Colesborne Park, Andoversford, R. S. O., Gloucestershire.

60 Yearling Ewes.

50 Two-year-old Ewes.

90 Older Ewes.

Flock Letter, E.

This Flock has been in existence on the Southbury Farm above 50 years, in the possession of H. Elwes, J. H. Elwes, A. Edmonds, and present owner.

FLETCHER, WILLIAM HINTON,

Shipton, Andoversford, R. S. O., Gloucestershire.

67 Yearling Ewes.

43 Two-year-old Ewes.

16 Older Ewes.

This Flock was established in 1866.

FOWLER, EDWARD POPE,

Aston Farm, Avening, Near Stroud.

95 Yearling Ewes.

90 Two-year-old Ewes.

115 Older Ewes.

This Flock was established in 1853.

FREEMAN, GEORGE,

Sherborne, Northleach, R. S. O., Gloucestershire.

90 Yearling Ewes.
100 Two-year-old Ewes.
180 Older Ewes.

Flock Letter, G. F.

This Flock was established by David Smith, Sherborne, about 1825, and was replenished in 1881 by Ewes (bred by the exors. of present owner's father) descended from the stock of Wm. Byan, Slaughter, and by Ewes bought at J. H. Pedley's sale.

GARDNER, THOMAS,

Clapton, Bourton-on-the-Water.

60 Yearling Ewes.
50 Two-year-old Ewes.
70 Older Ewes.

This Flock was established by Ewes bought from Compton Abdale 35 years ago.

GARNE, JOHN,

Filkins. Lechlade, Gloucestershire.

90 Yearling Ewes.
70 Two-year-old Ewes.
90 Older Ewes.

This Flock is descended from Ewes bred by William Garne, Aldsworth.

GARNE, ROBERT,

Aldsworth, Northleach, Gloucestershire.

130 Yearling Ewes.

110 Two-year-old Ewes.

184 Older Ewes.

Flock Letter, A.

This Flock came into the possession of present owner on the death of his father in 1857, by whom it was held from the year 1800, at which time it was a Flock of considerable note.

GILLETT, CHARLES,

Lower Haddon, Bampton, Faringdon.

60 Yearling Ewes.

160 Older Ewes.

This Flock has been established about 55 years.

GODWIN, JOHN JAMES,

Troy Farm, Somerton, Banbury, Oxon.

88 Yearling Ewes.

192 Older Ewes.

This Flock was established by William Godwin in 1818.

HANDY, AUBREY,

Coln St. Dennis, Northleach.

87 Yearling Ewes.

70 Two-year-old Ewes.

81 Older Ewes.

This Flock was established in the year 1884.

HANDY, EDWARD,

Shipton, near Andoversford.

74 Yearling Ewes.

87 Two year-old Ewes.

115 Older Ewes.

This Flock was established in 1876 by John Handy, and came into the possession of present owner in 1885.

HATHAWAY, ROBERT,

Cotswold Farm, Duntisbourne Abbotts, near Cirencester.

105 Yearling Ewes.

100 Two-year-old Ewes.

95 Older Ewes.

This Flock was established about 1852 from the Flocks of Barton, Coln Rogers, Henry Howell, and John Howell.

HOULTON, HENRY THOMAS,

Taynton, Burford, Oxon.

59 Yearling Ewes.

42 Two-year-old Ewes.

50 Older Ewes.

Flock Letter, H. H.

This Flock was established in 1871 by Ewes bought at Robt. Lane's sale, and of Thos. Craddock, Eastington, and of — Houlton, Ladbourn, and replenished in 1888 by Ewes bought at W. Lane's sale.

HOULTON, WILLIAM,

Broadfield Farm, Northleach, R. S. O.

26 Yearling Ewes.

39 Two-year-old Ewes.

37 Older Ewes.

Flock Letter, H. B.

This Flock was established by Ewes bought at the dispersion of the old Broadfield Flock in 1888.

HUCKRALE, JOHN EVANS,

Bruern Grange, Chipping Norton, Oxon.

65 Yearling Ewes.

63 Two-year-old Ewes.

72 Older Ewes.

This Flock was established in 1826.

HULBERT, T. R.,
North Cerney, Cirencester.

72 Yearling Ewes.

70 Two-year-old Ewes.

70 Older Ewes.

Flock Letter, H.

This Flock was established in 1874.

JACOBS, ROBERT.
Signett Hill, Burford, Oxon.

61 Yearling Ewes.

198 { Two-year-old Ewes
{ Older Ewes.

Flock Letter, J.

This Flock was established by Ewes bought in 1870 at — Faulkner's sale, Bury Barns, Burford, and replenished by Ewes bought of — Kerr, C. Pinnell, and W. Lane.

PORTER, THOMAS,
Baunton, Cirencester.

60 Yearling Ewes.

58 Two-year-old Ewes.

170 Older Ewes.

This Flock was established by the father of present owner almost 80 years ago.

POWELL, WALTER,

Upton Downs, Burford, Faringdon.

60 Yearling Ewes.

56 Two-year-old Ewes.

82 Older Ewes.

Flock Letter, P.

This Flock was established in 1876, and has been replenished by Ewes bought of Charles Pinnell and William Lane, Broadfield.

SMITH, JOHN WILLIAM,

Hill House, Shilton, Bampton, Oxon.

115 Yearling Ewes.

$204 \begin{cases} \text{Two-year-old Ewes.} \\ \text{Older Ewes.} \end{cases}$

This Flock was established in 1857.

SWANWICK, RUSSELL,

Royal Agricultural College Farm, Cirencester, Gloucestershire.

60 Yearling Ewes.

55 Two-year-old Ewes.

85 Older Ewes.

Flock Letter, R. S.

This Flock was established by Ewes bought at W. Hewer's Sale in 1869, and at R. Lane's Sale in 1871, and has been replenished by Ewes bought of William Lane.

TAYLOR, JAMES,

Rendcomb Park, Cirencester.

146 Yearling Ewes.

104 Two-year-old Ewes.

162 Older Ewes.

Flock Letter, R.

This Flock was established in 1882 by Ewes bought of — Brain , Shawswell Farm.

TAYLOR, JAMES,

Rendcomb Park, Cirencester.

77 Yearling Ewes.

80 Two-year-old Ewes.

109 Older Ewes.

Flock Letter, R.

This Flock was established in 1881 by Ewes bought at C. Newman's and Ely's sales.

TAYLOR, JAMES,

Rendcomb Park, Cirencester.

100 Yearling Ewes.

79 Two-year-old Ewes.

97 Older Ewes.

Flock Letter, R.

This Flock was established in 1881 by Ewes bought at C. Newman's and Ely's sales.

WAKEFIELD, JOHN PACKER,

Great Barrington, Burford, Oxon.

91 Yearling Ewes.

89 Two-year-old Ewes.

215 Older Ewes.

Flock Letter, I.

This Flock was established in 1834 by the purchase of Ewes from Charles Large, Broadwell.

WALKER, THOMAS,

Witney Street, Burford, Oxon.

70 Yearling Ewes.

60 Two-year-old Ewes.

74 Older Ewes.

Flock Letter, F.

This Flock was established by — Stratton about 1840, and came into possession of present owner in 1877.

WILLIAMS, OSWALD FIELD,

The Weir End, Ross, Herefordshire.

27 Yearling Ewes.

22 Two-year-old Ewes.

112 Older Ewes.

This Flock has been established about 32 years.

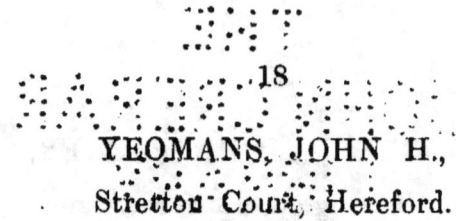

YEOMANS, JOHN H.,

Stretton Court, Hereford.

80 Yearling Ewes.

80 Two-year-old Ewes.

126 Older Ewes.

Flock Letter, Y.

This Flock is descended from 70 Ewes bought in 1850 of Wm. Garne, Kilkenny Farm, Bibury, and has been replenished by 5 Ewes bought at Wm. Hewer's sale in 1869, and by 15 bought at W. Lane's sale in 1888.

REGISTER OF RAMS,
1 to 403.

~~~~~~~~~o~~~~~~~

In the Registry of Cotswold Rams " *br.* " is used to signify " Breeder and Owner " where the animal has not changed ownership.

The other abbreviations used are " *ow.* " for Owner ; " *s.* " for Sire ; " *d.* " for Dam ; " *2nd d.* " for Grand-dam ; " *3rd d.* " for Great-grand-dam ; and " *4th d.* " for' Great-great-grand-dam. Also " R. A. S." for Royal Agricultural Society of England ; " B. & W. A. S." for Bath and West of England Agricultural Society ; " G. A. S." for Gloucestershire Agricultural Society ; " N. A. S." for Norfolk Agricultural Society ; " O. A. S." for Oxfordshire Agricultural Society ; " *r. n.* " for Reserve Number ; and " *h. c.* " for Highly Commended.

~~~~~~~~~~~~~~~~~~~~~~~~~~~

AARON—1.

Lambed 1889 ; *br.* R. Garne, *ow.* J. H. Yeomans, *s.* Excelsior 103, *d.* by Paragon 268.

ABBOT—2.

Lambed 1884 ; *br.* W. Lane, *ow.* J. H. Yeomans.

ABBOT—3.

Lambed 1886 ; *br.* R. Garne, *ow.* Thomas Brown, *s.* Ashbrook 16.

ABLINGTON—4.

Lambed 1887 ; *br.* Robert Garne, *s.* Norfolk.

ABLINGTON LYON—5.

Lambed 1878 ; *br.* Robert Garne, *s.* Young Lyon 400.

ADMIRAL—6.

Lambed 1878 ; *br.* Thomas Brown, *s.* Garne's Blackfoot 123.

ADRIAN—7.

Lambed 1889 ; *br.* R. Garne, *ow.* Thomas Brown & Son,
s. Master Charles 186.

AJAX—8.

Lambed 1887 ; *br.* R. Garne, *ow.* Thomas Brown & Son,
s. Master Charles 186.

ALAN—9.

Lambed 1889 ; *br.* Thomas Brown & Son, *s.* Ajax 8,
d. M 685 by Beacon 34, *2nd d.* by Admiral 6.

ALBERT—10.

Lambed 1890 ; *br.* Thomas Brown & Son, *s.* Ajax 8,
d. M 1284 by Abbot 3,
2nd d. M 280 by Gay Lad 126,
3rd d. by Dick Allen 87.

ALDSWORTH—11.

Lambed 1888 ; *br.* Robert Garne, *ow.* Charles Barton,
s. Garnet 124.

ALDSWORTH—12.

Lambed 1890 ; *br.* R. Garne, *ow.* G. Bagnall & Son,
s. Plymouth 274.

ALDSWORTH—13.

Lambed 1890 ; *br.* R. Garne, *ow.* T. R. Hulbert, *s.* Garnet 124.
d. by No. 2, 211, *2nd d.* by No. 2, 209,
3rd d. by The Marshal 368.

ANTIQUE—14.

Lambed 1882 ; *br.* W. Lane ; *ow.* J. H. Yeomans.

APPRENTICE—15.

Lambed 1882 ; *br.* W. Lane, *ow.* J. H. Yeomans.

ASHBROOK—16.

Lambed 1883 ; *br.* Robert Garne, *s.* Holwell.

ASTON—17.

Lambed 1888 ; *br.* E. Fowler. *ow.* J. H. Yeomans.

ATLAS—18.

Lambed 1889 ; *br.* R. Garne, *ow.* Thomas Brown & Son,
   s. 2nd Windsor 390.

AUBREY—19.

Lambed 1889 ; *br.* Thomas Brown & Son, *s.* Ajax 8,
   d. M 318 by Spot 348, *2nd d.* by Broadback 53.

BACHELOR—20.

Lambed 1890 ; *br.* C. Barton, *ow.* J. H. Yeomans, *s.* Noble 196,
   d. by Sir William 336.

BAMPTON—21.

Lambed about 1870 ; *br.* J. Gillett, *ow.* W. Lane, *s.* Broadfield.

BANGLE—22.

Lambed about 1860 ; *br.* W. Lane, *ow.* W. Hewer,
   s. Son of Bandy 342.

BARON—23.

Lambed 1890 ; *br.* C. Barton, *ow.* J. H. Yeomans, *s.* Captain 64,
   d. by B. Donner 31.

BARON FYFIELD—24.

Lambed 1890 ; *br.* Charles Barton, *ow.* Walter Powell,

s. No. 7 (Lane's), *d.* by Old Jack (Barton's).

BARTON—25.

Lambed 1890 ; *br.* T. Walker.

BATH AND WEST—26.

Prize, 1891, 1st B. & W. A. S.

Lambed 1890 ; *br.* Russell Swanwick, *s.* Windsor Reserve 388,

d. by the 50 Guinea Ram 365.

BATH—27.

Prize, 1891, 3rd B. & W. A. S.

Lambed 1890 ; *br.* Russell Swanwick, *s.* Royal Newcastle 303.

BAYARD—28.

Lambed 1885 ; *br.* Thomas Brown, *s.* Kent 163, *d.* by Thickset 374.

BANJO—29.

Lambed 1890 ; *br.* E. Brain, *ow.* Major the Hon. L. Byng.

B. D.—30.

Lambed 1887 ; *br.* W. Lane, *ow.* Robt. Garne, *s.* No. 9, 227.

B. DONNER—31.

Lambed 1884 ; *br.* Charles Barton, *s.* Donner 91,

d. by No. 6 (Lane's).

B. DONNER II.—32.

Lambed 1887 ; *br.* Charles Barton, *s.* B. Donner 31,

d. by O. Fowler 258.

B. DONNER III.—33.

Lambed 1890 ; *br.* Charles Barton, *s.* B. Donner II. 32,
d. by Broadfield 56.

BEACON—34.

Prize, 1884, 3rd R. A. S., Reading,

Lambed 1881 ; *br.* Thomas Brown, *s.* Example 102.

BEN—35.

Lambed 1891 ; *br.* Thos. Porter, *ow.* O. F. Williams.

BEN DONNER—36.

Lambed 1890 ; *br.* Charles Barton, *ow.* George Bagnall & Son,
s. B. Donner II. 32, *d.* by No. 7 (Lane's).

BERNARD—37.

Lambed 1886 ; *br.* Robert Garne, *s.* Peacemaker 269.

BERTRAM—38.

Lambed 1889 ; *br.* Robert Garne, *s.* Bernard 37.

BIG AYLMER—39.

Lambed 1888 ; *br.* Hugh Aylmer, *ow.* J. J. Godwin,
s. by No. 21 (Lane's).

BIG BEN—40.

Lambed 1888 ; *br.* Robert Jacobs, *s.* College Lad 74,
d. by Signett Donner, 322.

BIG BEN—41.

Lambed 1890 ; *br.* George Bagnall & Son, *s.* by Donner 91,
d. by Big Bill 44.

BIG NECK DONNER—42.

Lambed 1883 ; *br.* Charles Barton, *s.* Donner 91.

BIG NECKED DONNER II.—43.

Lambed 1888 ; *br.* C. Barton, *ow.* F. Craddock,

s. Big Neck Donner 42.

BIG BILL—44.

Prize, 1889, 1st R. A. S., Windsor.

Lambed 1887 ; *br.* G. Bagnall & Son.

BILL—45.

Lambed 1885 ; *br.* Charles Barton, *s.* Fowler, *d.* by No. 10, 230.

BILL—46.

Lambed 1890 ; *br.* Henry Clark, *ow.* C. E. Clark.

BLACKFOOT—47.

Lambed 1890 ; *br.* Charles Barton, *s.* Noble 196, *d.* by Prince 282.

BLUE LANE—48.

Lambed 1874 ; *br.* W. Lane, *ow.* Thomas Brown.

BLUNDERBUS—49.

Lambed 1889 ; *br.* Robert Garne, *ow.* John P. Wakefield,

s. Donner, 91.

BOB—50.

Lambed 1891 ; *br.* Thomas Porter, *ow.* O. F. Williams.

BOREAS—51.

Prize, 1874, 2nd N. A. S., Norwich.

Lambed 1873 ; *br.* Thomas Brown, *s.* Sir William, jun. 337.

d. by The Prince 370.

BOURTON—52.

Lambed 1889 ; *br.* Charles Barton, *ow.* Edward Brain,
s. Young Ben.

BROADBACK—53.

Lambed 1879 ; *br.* Thomas Brown, *s.* Broadfield 54.

BROADFIELD—54.

Lambed 1877 ; *br.* William Lane, *ow.* Thomas Brown.

BROADFIELD—55.

Br. W. Lane, *ow.* G. Bagnall & Son.

BROADFIELD—56.

Lambed 1886 ; *br.* W. Lane, *ow.* Charles Barton.

BROADFIELD—57.

Lambed 1888 ; *br.* W. Lane, *ow.* J. H. Yeomans.

BROADFIELD STAMP—58.

Lambed 1887 ; *br.* William Lane, *ow.* Geo. Freeman,
d. by Jumbo, 154.

BUMPER—59.

Lambed 1890 ; *br.* E. Brain, *ow.* Major the Hon. L. Byng.

BURSAR—60.

Prize, 1889, 3rd R. A. S., Windsor.

Lambed 1888 ; *br.* Thomas Brown & Son, *s.* Bernard 37,
d. M 294 by Spot 348,
2nd d. by Royal Kilburn 302.

CADET—61.

Lambed 1887 ; *br.* Thomas Brown, *ow.* Robert Garne,
s. Slasher 339, *d.* by No. 20, 246,
2nd d. by Ensign 99.

CADET—62.

Lambed 1887 ; *br.* Thomas Brown, *s.* Cavalier 66,
d. by Dick Allen 87.

CAPTAIN—63.

Lambed 1872 ; *br.* Robert Garne, *s.* Magdala 183.

CAPTAIN—64.

Lambed 1886 ; *br.* Charles Barton, *s.* O. Fowler 258.

CAPTAIN JUMBO—65.

Lambed 1885 ; *br.* Robert Garne, *s.* No. 5, 221.

CAVALIER—66.

1886, *h. c.* R. A. S., Norwich.

Lambed 1885 ; *br.* Thomas Brown, *s.* Surprise 353,
d. by Sultan 352.

CHAMPION—67.

Lambed 1879 ; *br.* Charles Barton, *s.* No. 20, 245.

CHARLIE—68.

Lambed 1889 ; *br.* Robert Jacobs, *s.* Signett Donner 322,
d. by Duke of Carlisle 93.

CICETER—69.

Lambed 1890 ; *br.* H. J. Elwes, *s.* No. 5 (Swanwick's)

CITADEL—70.

Lambed 1885 ; *br.* Thomas Brown, *s.* Surprise 353,
d. by Boreas 51.

CLAUDIAN—71.

Lambed 1887 ; *br.* Thomas Brown, *s.* Cavalier 66,
d. by Example 102.

COLIN—72.

Lambed 1887 ; *br.* Thomas Brown, *s.* Cavalier 66, *d.* by Sultan 352.

COLLEGIAN—73.

Lambed 1885; *br.* R. Swanwick, *ow.* Robert Jacobs, *s.* Young Tim.

COLLEGE LAD—74.

Lambed 1887 ; *br.* Robert Jacobs, *s.* Collegian 73,
d. by Signett Donner 32?.

COMET—75.

Lambed 1885 ; *br.* Thomas Brown, *s.* Surprise 353,
d. by Broadfield 54.

COMMODORE—76.

Lambed 1880; *br.* Thomas Brown, *s.* K. B. 159.

CONRAD—77.

Lambed 1885 ; *br.* Thomas Brown, *s.* Surprise 353,
d. by Broadfield 54.

CONSTITUTION—78.

Lambed 1879 ; *br.* Robert Garne, *s.* Experience 106,
d. by The Squire 371.

CONSTITUTION—79.

Lambed 1890 ; *br.* C. Barton, *ow.* Henry Clark, *s.* Noble 196,
d. by Old Jack.

COTSWOLD KING—80.

Lambed 1864 ; *br.* W. Hewer, *ow.* W. Lane.

CURLY—81.

Lambed 1889 ; *br.* Robert Garne, *s.* Cadet 61.

CURLY—82.

Lambed 1890 ; *br.* Charles Barton, *s.* No. 30, 251, *d.* by Noble 196.

CURLY—83.

Lambed 1890 ; *br.* T. Walker.

CURLY LOCKS—84.

Lambed 1890 ; *br.* Russell Swanwick, *s.* Windsor Reserve 388.

DANDY—85.

Lambed 1890 ; *br.* Charles Barton, *ow.* Thomas Brown & Son,
s. No. 7 (Lane's), *d.* by Jack 148.

D. D.—86.

Lambed 1873 ; *br.* Thomas Brown, *s.* the Doctor 361.

DICK ALLEN—87.

Prize, 1879, 1st R. A. S., Kilburn.

Lambed 1876 ; *br.* Thomas Brown, *s.* No. 4, 219.

DEREHAM—88

Lambed 1873 ; *br.* H. Aylmer, *ow.* Robt. Garne, *s.* No. 20, 244.

DINGER—89.

Lambed 1877 ; *br.* Charles Barton,

s. No. 5 (C. Barton's, Coln Deans).

DINGER—90.

Lambed about 1878 ; *br.* C. Barton, *s.* Old Sam 262.

DONNER—91.

Lambed 1881 ; *br.* Charles Barton, *s.* Champion 67.

DONNER II.—92.

Lambed 1884 ; *br.* C. Barton, *ow.* Robt. Garne, *s.* Donner 91,
d. by Old Forty.

DUKE OF CARLISLE—93.

Prizes, 1880 $\begin{cases} \text{1st R. A. S., Carlisle.} \\ \text{1st B. \& W. A. S. ; G. A. S. ; and O. A. S.} \end{cases}$

Lambed 1879 ; *br.* Robert Jacobs, *s.* Sir Arthur 325.

DUKE OF YORK—94.

Prize, 1883, 2nd R. A. S., York.

Lambed 1882 ; *br.* Thomas Brown, *s.* Experience 106.

EASTINGTON—95.

Lambed 1888 ; *br.* F. Craddock, *ow.* John P. Wakefield.

ECLIPSE—96.

Prize, 1886, 2nd R. A. S., Norwich

Lambed 1884 ; *br.* Thomas Brown, *s.* Example 102.

EIGHTY GUINEA—97.

Lambed 1869 ; *br.* William Lane, *ow.* Russell Swanwick,

s. Cotswold King 80, *d.* by Young Sweetbriar,

2nd d. by Tiny 379.

EMPEROR—98

Lambed 1890 ; *br.* Robert Garne, *ow.* Robt. Jacobs,
s. Windsor 2nd 390, *d.* by Exchange 104.

ENSIGN—99.

Lambed 1875 ; *br.* Thomas Brown, *s.* The Marshal 367.

ENTERPRISE—100.

Lambed 1881 : *br.* Robert Garne, *s.* Example 102.

ERIC—101.

Lambed 1886 ; *br.* Thomas Brown, *s.* Excelsior 103,
d. by Sultan 352.

EXAMPLE—102.

Lambed 1879 ; *br.* Robert Garne, *s.* Experience 106.

EXCELSIOR—103.

Lambed 1884 ; *br.* Robt. Garne, *s.* Enterprise 100.

EXCHANGE—104.

Lambed 1881 ; *br.* T. Brown, *ow.* Robt. Garne,
s. Marham Blackfoot 185.

EXPERIMENT—105.

Lambed 1875 ; *br.* Robt. Garne, *s.* Dereham 88.

EXPERIENCE—106.

Lambed 1877 ; *br.* Robt. Garne, *s.* Experiment 105.

FASHION—107.

Lambed 1889 ; *br.* Thomas Brown & Son, *s.* Fyfield 122,
d. M 393 by Example 102, *2nd d.* by Reserve 294.

FATHER CHRISTMAS—108.

Lambed 1888 ; *br.* Robert Garne, *ow.* John P. Wakefield,

    s. Donner 2nd, 92.

FAVORITE—109.

1886, *h. c.* R. A. S., Norwich.

Lambed 1884 ; *br.* Thomas Brown, *s.* Haylock 146,

    d. by Grey Nobleman 142.

FELIX—110.

Lambed 1890 ; *br.* Thomas Brown & Son, *s.* Forester 113,

    d. M 1301 by Bernard 37,

    2nd d. M 496 by Example 102,

    3rd d. by Reserve 294.

FILKINS—111.

Lambed 1889 ; *br.* John Garne, *s.* Donner 91.

FINCHAM—112.

Lambed 1890 ; *br.* J. B. Aylmer, *s.* Fat Back.

FORESTER—113.

Lambed 1887 ; *br.* Thomas Brown & Son, *s.* Favorite 109,

    d. M 275 by Duke of York 94,

    2nd d. by No. 1, 205.

FORESTER—114.

Lambed 1889 ; *br.* J. H. Yeomans, *s.* Fyfield 121.

FORTY GREY—115.

Lambed 1876 ; *br.* Charles Barton.

FORTY WHITE—116.

Lambed 1876; *br.* Charles Barton.

FOWLER—117.

Lambed 1890; *br.* J. H. Yeomans, *s.* Aston 17, *d.* by Abbot 2.

FRAMPTON—118.

Lambed 1890; *br.* H. Clark, *s.* Captain.

FREEMAN—119.

Lambed 1889; *br.* J. H. Yeomans, *s.* Fyfield 121.

FRED—120.

Lambed 1888; *br.* Robert Garne, *s* Peacemaker 269,
d. by Timothy 377.

FYFIELD—121.

Lambed 1886; *br.* C. Barton, *ow.* J. H. Yeomans, *s.* Old Jack.

FYFIELD—122.

Lambed 1887; *br.* Charles Barton, *ow.* Thomas Brown & Son,
s. Jack 148, *d.* by Big Neck Donner 42.

GARNE'S BLACKFOOT—123.

Lambed 1876; *br.* R. Garne, *ow.* Thomas Brown,
s. The Second Squire 372.

GARNET—124.

Lambed 1884; *br.* Robert Garne, *s.* No. 2, 211.

GARNET II.—125.

Lambed 1889; *br.* Robert Garne, *s.* Garnet 124.

GAY LAD—126.

Lambed 1880 ; *br.* Thomas Brown, *s.* Royal Kilburn 302.

GENERAL—127.

Prize, 1863, 2nd R. A. S., Worcester.

Lambed 1862 ; *br.* R. Garne, *ow.* Thomas Brown.

GENERAL—128.

Prize, 1873, 1st R. A. S., Hull.

Lambed 1872; *br.* Thomas Brown, *s.* No. 7, 223, *d.* by General 127.

GENERAL—129.

Prizes $\begin{cases} \text{1876, 1st R. A. S., Birmingham.} \\ \text{1878, 2nd R. A. S., Bristol.} \end{cases}$

Lambed 1875 ; *br.* Thomas Brown, *s.* General 128.

GIANT—130.

Lambed 1888 ; *br.* Hugh Aylmer, *ow.* Russell Swanwick,

s. Perfection.

GOOD BOY—131.

Lambed 1883 ; *br.* Robert Garne, *s.* No. 13, 237.

GOOD HEAD—132.

Lambed 1886 ; *br.* Charles Barton, *ow.* C. Gillett.

GOODMAN—133.

Lambed 1885 ; *br.* R. Jacobs, *ow.* J. J. Godwin,

s. Little Norfolk 178.

GRAMP—134.

Lambed 1880 ; *br.* Charles Barton, *s.* No. 10, 230.

GRAMP II.—135.

Lambed 1884 ; *br.* Charles Barton, *s.* Gramp 134.

GRANITE—136.

Lambed 1886 ; *br.* Thomas Brown, *s.* Garnet 124, *d.* by K. B. 159.

GREAT LONG JOHN—137.

Lambed 1887 ; *br.* Robert Jacobs, *ow.* C. Gillett,
s. Long John 179.

GRENADIER—138.

Lambed 1888 ; *br.* Thomas Brown & Son, *s.* Granite 136,
d. M 342 by Beacon 34,
2nd d. by Royal Kilburn 302.

GREY FACE—139.

Lambed 1884 ; *br.* Robert Garne, *s.* Constitution 78.

GREY FACE II.—140.

Lambed 1888 ; *br.* Hugh Aylmer, *s.* Grey Face 139.

GREY LANE—141.

Lambed 1874 ; *br.* William Lane, *ow.* Thomas Brown.

GREY NOBLEMAN—142.

Lambed 1875 ; *br.* T. Brown, *s.* Lane's Two-shear Sheep 167.

GREY SON OF BROADFIELD—143.

Lambed 1879 ; *br.* Thomas Brown, *s.* Broadfield 54.

GREY SON OF LITTLE DOCTOR—144.

Lambed 1877 ; *br.* Thomas Brown, *s.* Little Doctor 177.

HARDY TOM—145.

Lambed 1887 ; *br.* H. Clark, *s.* Captain.

HAYLOCK—146.

Lambed 1881 ; *br.* Thomas Brown, *s.* No. 2, 210.

HECTOR—147.

Lambed 1889 ; *br.* Thomas Brown & Son, *s.* Fyfield 122,

d. M 1053 by Paragon 268,

2nd d. M 288 by Commodore 76,

3rd d. by Broadfield 54.

JACK—148.

Lambed 1881 ; *br.* Charles Barton, *s.* No. 4, 220.

JACK—149.

Lambed 1889 ; *br.* J. P. Wakefield, *ow.* Edward Brain.

JACK—150.

Lambed 1890 ; *br.* O. F. Williams.

JACOB—151.

Lambed 1886 ; *br.* R. Jacobs, *ow.* J. H. Yeomans.

JOE—152.

Lambed 1889 ; *br.* Henry Clark, *ow.* C. E. Clark, *s.* Old Joe.

JUBILEE—153.

Lambed 1887 ; *br.* Charles Barton, *ow.* John Garne.

JUMBO—154.

Lambed 1880 ; *br.* R. Garne, *ow.* W. Lane, *s.* Young Captain 394,

d. by Crown Prince.

JUMBO—155.

Prize, 1888, 1st R. A. S., Nottingham.

Lambed 1886; *br.* G. Bagnall & Son.

JUMBO—156.

Lambed 1889 ; *br.* W. J. Smith.

JUMBO—157.

Lambed 1890 ; *br.* H. Clark, *s.* Captain.

JUMBO JUNIOR—158.

Lambed 1890 ; *br.* W. J. Smith.

K. B.—159.

Lambed 1878 ; *br.* Thomas Brown, *s.* Garne's Blackfoot 123.

KILKENNY—160.

Lambed 1890 ; *br.* T. Gillett, *ow* J. J. Godwin.

KING—161.

Lambed 1883 ; *br.* Charles Barton, *s.* Prince 282, *d.* by No. 4, 220.

KING B.—162.

Lambed 1887 ; *br.* Charles Barton, *ow.* Robt. Jacobs,
s. Old Jack, *d.* by Big Neck Donner 42.

KENT—163.

Lambed 1881 ; *br.* Thomas Brown, *s.* Marham Blackfoot 185.

LANCER—164.

Lambed 1886 ; *br.* Wm. Lane, *ow.* Thomas Brown.

LANE'S CHOICE—165.

Lambed 1888 ; *br.* R. Swanwick, *ow.* H. J. Elwes.

LANE'S OLD SHEEP—166.

Lambed 1873 ; *br.* Wm. Lane, *ow.* Thomas Brown.

LANE'S TWO-SHEAR SHEEP—167.

Lambed 1872 ; *br.* Wm. Lane.

LANDSMAN—168.

Lambed 1879 ; *br.* W. Lane, *ow.* J. H. Yeomans.

LANE JUNIOR—169.

Lambed 1889 ; *br.* Hugh Aylmer, *s.* Old Joe.

LAST OF ALL—170.

Lambed 1885 ; *br.* Wm. Lane, *ow.* John P. Wakefield.

LATIMER—171.

Lambed 1888 ; *br.* W. Lane, *ow.* J. H. Yeomans.

LATITUDE—172.

Lambed 1879 ; *br.* W. Lane, *ow.* J. H. Yeomans.

LEONARD—173.

Lambed 1887 ; *br.* Wm. Lane, *ow.* Thomas Brown & Son.

LESLIE—174.

Lambed 1889 ; *br.* Thomas Brown & Son, *s.* Leonard 173.

d. M 938 by Favorite 109, *2nd d.* by Example 102, *3rd d.* Old Grey by No. 4, 219.

LION—175.

Lambed 1883 ; *br*. Wm. Lane, *ow*. Thomas Brown.

LITTLE AYLMER—176

Lambed 1888 ; *br*. Hugh Aylmer, *ow*. J. J. Godwin.

LITTLE DOCTOR—177.

Lambed 1875 ; *br*. Thomas Brown, *s*. D. D. 86.

LITTLE NORFOLK—178.

Lambed 1882 ; *br*. Hugh Aylmer, *ow*. Robt. Jacobs,
s. No. 12 (Lane's).

LONG JOHN—179

Lambed 1881 ; *br*. C. E. Clark, *ow*. Robt. Jacobs.

LORD CHARLES—180.

Lambed 1879 ; *br*. Chas. Barton, *ow*. Robt. Jacobs, *s*. No. 20, 245.

LORD LYON—181.

Lambed 1865 ; *br*. R. Garne, *ow*. Thomas Brown.

LUCKLESS—182.

Lambed 1868 ; *br*. Robt. Garne, *s*. Old Sierford 263.

MAGDALA—183.

Lambed 1866 ; *br*. Robert Garne, *s*. the Big Necked Sheep 359.

MAJOR—184.

Lambed 1888 ; *br*. H. J. Clark, *ow*. C. E. Clark.

MARHAM BLACKFOOT—185.

Prizes, 1878 { 2nd N. A. S., North Walsham.
 r. s. R. A. S., Bristol.

Lambed 1877 ; *br.* Thomas Brown, *s.* Lane's Old Sheep.

MASTER CHARLES—186.

Lambed 1885 ; *br.* C. Barton, *ow.* Robert Garne, *s.* Donner 91.

d. by The Old Grey.

MASTER DONNER—187.

Lambed 1890 ; *br.* R. Garne, *ow.* George Freeman,

s. Donner 2nd 92.

MILLER—188.

Lambed 1887 ; *br.* William Lane, *ow.* Walter Powell.

MODEL—189.

Lambed 1888 ; *br.* T. Brown, *ow.* Robert Garne, *s.* Bernard 37,

d. by Excelsior 103.

MORTON—190.

Lambed 1871 ; *br.* Thomas Brown, *s.* Senior Wrangler 313.

MOSES—191.

Lambed 1877 ; *br.* Chas. Barton, *s.* No. 5, Barton's Coln Deans.

MOSES—192.

Lambed 1889 ; *br.* Ed. Brain.

MOSES—193.

Lambed 1889 ; *br.* Robert Garne, *ow.* J. H. Yeomans,

s. Donner 2nd 92.

MULTUM IN PARVO—194.

Lambed 1889 ; *br.* Robert Garne, *ow.* W. J. Elwes, *s.* Cadet 61.

NELSON—195.

Lambed 1889 ; *br.* Charles Barton, *s.* B. Donner 31,
d. by No. 30, 251.

NOBLE—196.

Lambed 1886 ; *br.* Charles Barton, *s.* No. 1, 206, *d.* by Donner 91.

NOBLE—197.

Lambed 1889 ; *br.* Charles Barton, *ow.* George Bagnall & Son,
s. Noble 196.

NOBLE BEN—198.

Lambed 1890 ; *br.* Charles Barton, *ow.* Robert Jacobs,
s. Noble 196, *d.* by Ben Donner.

NOBLEMAN—199.

Prize, 1876, 2nd R. A. S., Birmingham.
Lambed 1875 ; *br.* Thomas Brown,
s. Lane's Two-shear Sheep 167.

NO NAME—200.

Lambed 1868 ; *br.* Robert Lane, *ow.* Robert Garne,
s. Cotswold King 80.

NO No.—201.

Lambed 1873 ; *br.* J. H. Pedley, *ow.* Thomas Brown.

NORTHAMPTON FIRST—202.

Lambed 1886 ; *br.* Russell Swanwick, *s.* No. 9, 228.

No. 1—203.

Lambed 1875 ; *br.* Robert Garne, *s.* Tiny 379.

No. 1—204.

Lambed 1876 ; *br.* Robert Garne.

No. 1—205.

Lambed 1876 ; *br.* William Lane.

No. 1—206.

Lambed 1884 ; *br.* W. Lane, *s.* Jumbo 154.

No. 1—207.

Lambed 1887 ; *br.* T. Porter.

No. 1—208.

Lambed 1889 ; *br.* Edward Brain.

No. 2—209.

Lambed 1878 ; *br.* Robert Garne, *s.* The Marshal 368.

No. 2—210.

Lambed 1879 ; *br.* Wm. Lane.

No. 2—211.

Lambed 1881 ; *br.* Robert Garne, *s.* No. 2, 209.

No. 2—212.

Lambed 1885 ; *br.* Thos. Brown, *s.* Kent 163,
d. by White Nobleman 385.

No. 2—213.

Lambed 1887 ; *br.* Mr. C. Barton, *ow.* George Beak, *s.* Old King,
d. by Sir William 336.

No. 2—214.

Lambed 1889 ; *br.* Ed. Brain.

No. 3—215.

Lambed 1872 ; *br.* Robert Garne.

No. 3—216.

Lambed 1887 ; *br.* T. Porter.

No. 3—217.

Lambed 1890 ; *br.* C. Barton, *ow.* George Beak,
     s. Captain 64, *d.* by Old Blackfoot 260.

No. 3, 2ND—218.

Lambed 1874 ; *br.* Thomas Brown, *s.* No. 3, 215.

No. 4—219.

Lambed 1874 ; *br.* J. H. Pedley, *ow.* Thomas Brown.

No. 4—220.

Lambed 1879 ; *br.* W. Lane.

No. 5—221.

Lambed 1883 ; *br.* W. Lane, *ow.* Robert Garne, *s.* Jumbo 154.

No. 5—222.

Lambed 1873 ; *br.* Thomas Brown, *s.* The Doctor 361.

No. 7—223.

Lambed 1870 ; *br.* Robert Garne, *ow.* Thomas Brown.

No. 7—224.

Lambed 1888 ; *br.* C. Barton, *ow.* George Beak, *s.* T. Donner 357.

No. 8—225.

Lambed 1878 ; *br.* Robert Garne, *ow.* Thomas Brown.

No. 9—226.

Lambed 1881 ; *br.* Wm. Lane, *ow.* Thomas Brown.

No. 9—227.

Lambed 1882 ; *br.* Robert Garne, *ow.* W. Lane,
s. Constitution 78.

No. 9—228.

Lambed 1883 ; *br.* Charles Barton, *ow.* Russell Swanwick,
s. Blackfoot (Barton's).

No. 9—229.

Lambed 1889 ; *br.* C. Barton, *ow.* T. Porter, *s.* Old Bill,
d. by No. 17, 242.

No. 10—230.

Lambed 1878 ; *br.* W. Lane, *ow.* Charles Barton,
s. No. 10 (Barton's), *d.* by Tiny 379.

No. 10—231.

Lambed 1885 ; *br.* Robert Garne, *ow.* Russell Swanwick,
s. Paragon 268.

No. 10—232.

Lambed 1889 ; *br.* Robert Garne, *ow.* J. J. Godwin, *s.* Bernard 37.

No. 10—233.

Lambed 1890 ; *br.* Robert Garne, *ow.* T Walker,

s. Donner 2nd 92.

No. 12—234

Lambed 1876 ; *br.* Robert Garne, *ow.* T. Brown,

s. The Second Squire 372.

No. 12—235.

Lambed 1889 ; *br.* C. Barton, *ow.* George Beak, *s.* Old King.

No. 12—236.

Lambed 1890 ; *br.* C. Barton, *ow.* H. J. Elwes, *s.* No. 7 (Lane's)

d. by T. Donner 357.

No. 13—237.

Lambed 1881 ; *br.* C. Barton, *ow.* Robert Garne, *s.* Dinger 90.

No. 13—238.

Lambed 1884 ; *br.* R. Garne, *ow.* Thomas Brown.

No. 13—239.

Lambed 1888 ; *br.* Robert Garne, *ow.* T. Porter, *s.* Professor 288.

No. 14—240.

Lambed 1887 ; *br.* William Lane, *ow.* Russell Swanwick,

s. B. D. (Lane's).

No. 15—241.

Lambed 1867 ; *br.* G. Fletcher, *ow.* W. Lane, *s.* No. 40, 253

No. 17—242.

Lambed 1879 ; *br.* W. Lane, *ow.* Charles Barton.

No. 19—243.

Lambed 1872 ; *br.* Robert Garne, *s.* The Squire 371.

No. 20—244.

Lambed 1871 ; *br.* W. Lane, *ow.* H. Aylmer, *s.* No. 15, 241.

No. 20—245.

Lambed 1877 ; *br.* E. Handy, *ow.* Charles Barton.

No. 20—246.

Lambed 1880 ; *br.* Thomas Brown, *s.* Royal Kilburn 302.

No. 20—247.

Lambed 1883 ; *br.* W. Lane, *ow.* Robert Garne, *s.* Old Joe.

No. 20—248.

Lambed 1889 ; *br.* Robert Jacobs, *ow.* J. J. Godwin, *s.* Spot 349.

No. 20—249.

Lambed 1890 ; *br.* C. Barton, *ow.* George Beak, *s.* Noble 196,
d. by Ben Donner.

No. 20—250.

Lambed 1890 ; *br.* Russell Swanwick, *ow.* T. Porter.

No. 30—251.

Lambed 1885 ; *br.* W. Lane, *ow.* Charles Barton.

No. 39—252.

Lambed 1887 ; *br* William Lane, *ow.* Russell Swanwick,
s. Donner 91.

No. 40—253.

Lambed 1865 ; *br.* W. Lane, *ow.* G. Fletcher, *s.* Old Ought 261.

No. 40—254.

Lambed 1885 ; *br.* R. Garne, *ow.* George Freeman, *s.* No. 20, 247.

No. 40—255.

Lambed 1887 ; *br.* William Lane, *ow.* Russell Swanwick,
s. Donner 91.

No. 42—256.

Lambed 1877 ; *br.* William Lane, *ow.* Thomas Brown.

No. 62—257.

Lambed 1889 ; *br.* Hugh Aylmer, *ow.* Russell Swanwick,
s. Broadfield (Aylmer's).

O. FOWLER—258.

Lambed 1881 ; *br.* Edward Fowler, *ow.* Charles Barton.

OLD BARTON—259.

Lambed 1885 ; *br.* Charles Barton, *ow.* J. J. Godwin,
s. O. Fowler 258, *d.* by No. 10, 230.

OLD BLACKFOOT—260.

Lambed 1879 ; *br.* Charles Barton, *s.* No. 20, 245.

OLD OUGHT—261.

Lambed about 1860 ; *br.* Robert Lane, *s.* Hewer's Old Ought.

OLD SAM—262.

Lambed about 1876 ; *br.* C. Barton.

OLD SIERFORD—263.

Lambed about 1864 ; *br.* E. Handy, *ow.* Robert Garne,
s. Young Substance 403.

OLIVER—264.

Lambed 1886 ; *br.* Thomas Brown, *s.* No. 13, 238,
d. M 62 by Commodore 76, *2nd d.* by General 129.

OSCAR—265.

Lambed 1887 ; *br.* Robert Garne, *ow.* Thomas Brown & Son,
s. No. 5 (Lane's).

OXFORD—266.

Lambed 1889 ; *br.* Russell Swanwick, *s.* No. 39, 252,
d. by the 50 Guinea Ram 365.

PALE FACE—267.

Lambed 1888 ; *br.* Robert Garne ; *s.* Professor 288.

PARAGON—268.

Lambed 1883 ; *br.* Thomas Brown, *s.* Spot 348, *d.* by Prince 281.

PEACEMAKER—269.

Lambed 1884 ; *br.* Robert Garne ; *s.* Exchange 104,
d. Countess by No. 2, 209.

PERCY—270.

Lambed 1885 ; *br.* R. Garne, *ow.* Thomas Brown, *s.* Paragon 268.

PERFECTION—271.

Lambed 1890 ; *br.* C. Barton, *ow.* William Houlton,
   s. B. Donner 2nd, 32 ; *d.* by Captain 64.

PERFECTION—272.

Lambed 1888 ; *br.* Robert Garne ; *ow.* John Garne,
   s. Peacemaker 269.

PILOT—273.

Lambed 1888 ; *br.* Thomas Brown & Son, *s.* Percy 270,
   d. M 35 by Royal Kilburn 302,
   2nd d. by No. 3, 2nd, 218.

PLYMOUTH—274.

Prize, 1890, 1st R. A. S., Plymouth.
Lambed 1888 ; *br.* Robert Garne, *s.* Eric 101, *d.* by Excelsior 103.

PORTER'S LONGBACK—275.

Lambed 1890 ; *br.* T. Porter, *ow.* H. J. Elwes.

PORTRAIT—276.

Lambed 1890 ; *br.* Robert Garne ; *s.* Pale Face 267.

PRELATE—277.

Lambed 1888 ; *br.* Thomas Brown & Son, *s.* Primate 278,
   d. M 465 by Lion 175, *2nd d.* by No. 42, 256.

PRIMATE—278.

Lambed 1886 ; *br.* Thomas Brown, *s.* Spot 348,
   d. by Broadback 53.

PRINCE—279.

Prize, 1875, 1st R. A. S., Taunton.

Lambed 1874 ; *br.* Thomas Brown, *s.* Prince III. 284.

PRINCE—280.

Lambed 1879 ; *br.* Charles Barton, *s.* Dinger 89.

PRINCE—281.

Lambed 1889 ; *br.* Charles Barton, *ow.* William Houlton,

s. Captain 64, *d.* by B. Donner 31.

PRINCE—282.

Lambed 1889 ; *br.* Robert Garne, *s.* Donner II. 92,

d. by Jumbo 154.

PRINCE II.—283.

Prize, 1871, 2nd R. A. S., Wolverhampton.

Lambed 1870 ; *br.* Thomas Brown, *s.* The Prince 370.

PRINCE III.—284.

Lambed 1872 ; *br.* Thomas Brown, *s.* Prince II. 283.

PRIOR—285.

Lambed 1888 ; *br.* Thomas Brown & Son, *s.* Paragon 268,

d. M 831 by Garnet 124,

2nd d. M 142 by Quality 289,

3rd d. by Grey Nobleman 142.

PRIVATE—286.

Lambed 1887 ; *br.* J. J. Godwin, *ow.* Robert Jacobs.

PRIVATEER—287.

Lambed 1889 ; *br.* Robert Garne, *s.* No. 19, 243.

PROFESSOR—288.

Lambed 1885 ; *br.* Robert Garne, *s.* No. 20, 247.

QUALITY—289.

Lambed 1879 ; *br.* Robert Garne, *s.* Royal Kilburn 302.

QUALITY—290.

Lambed 1890 ; *br.* Robert Jacobs, *s.* Russett 310,
d. by Little Norfolk 178.

QUALITY—291.

Lambed 1890 ; *br.* George Freeman, *s.* Broadfield Stamp 58,
d. by Sherborne 314.

REAR ADMIRAL—292.

Lambed 1882 ; *br.* Thomas Brown, *s.* Admiral 6.

REMUS—293.

Lambed 1887 ; *br.* Charles Barton, *ow.* Henry Thomas Houlton,
s. B. Donner 31.

RESERVE—294.

1879, *r. n.* R. A. S., Kilburn.
Lambed 1878 ; *br.* Thomas Brown, *s.* Nobleman 199.

ROLAND—295.

Lambed 1890 ; *br.* T. R. Hulbert, *s.* B. S.

ROMULUS—296.

Lambed 1887 ; *br.* Robert Garne, *ow.* Henry Thomas Houlton,
s. No. 2, 212.

RONALD—297.

Lambed 1883 ; *br.* Thomas Brown, *s.* Reserve 294,
d. by Boreas 51.

ROUGH DONNER—298.

Lambed 1883 ; *br.* Charles Barton, *s.* Donner 91.

ROUGH HEAD—299.

Lambed 1887 ; *br.* J. J. Godwin, *s.* Goodman 133.

ROYAL DERBY—300.

Prize, 1881, 2nd R. A. S., Derby.

Lambed 1880 ; *br.* Thomas Brown, *s.* Reserve 294.

ROYAL DONCASTER—301.

Prize, 1891, 1st R. A. S., Doncaster.

Lambed 1890 ; *br.* Robert Garne, *s.* B. D. 30,
d. Old Bet by Experiment 105.

ROYAL KILBURN—302.

Prize, 1879, 1st R. A. S., Kilburn,

Lambed 1878 ; *br.* Thomas Brown, *s.* Garne's Blackfoot 123.

ROYAL NEWCASTLE—303.

Prize, 1887, 1st R. A. S., Newcastle.

Lambed 1885 ; *br.* Russell Swanwick, *s.* the 50 Guinea Ram 365.

ROYAL PLYMOUTH—304.

Prize, 1890, 1st R. A. S., Plymouth.

Lambed 1889 ; *br.* Russell Swanwick, *s.* No. 39, 252.

ROYAL READING—305.

Prize, 1882, 1st R. A. S., Reading.

Lambed 1881 ; *br.* Robert Jacobs, *s.* Lord Charles 180.

ROYAL WINDSOR—306.

Prize, 1889, 2nd R. A. S., Windsor.

Lambed 1888 ; *br.* Russell Swanwick, *ow.* Hugh Aylmer.

ROUGH—307.

Lambed 1888 ; *br.* Charles Barton, *s.* Broadfield 56,
d. by Gramp 134.

RUMPPY—308.

Lambed 1885 ; *br.* T. Porter, *ow.* T. R. Hulbert.

RUPERT—309.

Lambed 1885 ; *br.* Thomas Brown, *s.* Lion 175,
d. by Reserve 294.

RUSSETT—310.

Lambed 1884 ; *br.* Robert Jacobs, *s.* Duke of Carlisle 93,
d. by Lord Charles 180.

SAMSON—311.

Lambed 1890 ; *br.* Charles Barton, *s.* Broadfield 56,
d. by Sir William 336.

SANFOIN—312.

Lambed 1889; *br.* Robert Jacobs, *ow.* J. J. Godwin,

s. Signett Donner 322.

SENIOR WRANGLER—313.

Prize, 1870, 1st R. A. S., Oxford.

Lambed 1869; *br.* Thomas Brown, *s.* Lord Lyon 181,

d. by Sir James 330.

SHERBORNE—314.

Lambed 1886; *br.* Robert Garne, *ow.* G. Freeman, *s.* Quality 289.

SHYLOCK—315.

Lambed 1885; *br.* Thomas Brown, *s.* Singleton 324,

d. by No. 8, 225.

SIGNETT—316.

Lambed 1888; *br.* Robert Jacobs, *ow.* Robert Garne,

s. Little Norfolk 178, *d.* by Royal Reading 305.

SIGNETT II.—317.

Lambed 1888; *br.* R. Jacobs, *ow.* George Bagnall & Son,

s. Little Norfolk 178.

SIGNETT III.—318.

Lambed 1889; *br.* R. Jacobs, *ow.* George Bagnall,

s. College Lad 74.

SIGNETT IV.—319.

Lambed 1890; *br.* R. Jacobs, *ow.* George Bagnall & Son,

s. Young Lane 399.

SIGNETT—320.

Lambed 1889 ; *br*. Robert Jacobs, *ow*. Walter Powell,
s. Signett Donner 322.

SIGNETT—321.

Lambed 1890 ; *br*. Robert Jacobs, *ow*. Charles Barton.

SIGNETT DONNER—322.

Lambed 1884 ; *br*. Charles Barton, *ow*. Robert Jacobs,
s. Donner 91, *d*. by Old Forty Grey 115.

SILVIO—323.

Lambed 1890 ; *br*. C. Barton, *ow*. William Houlton,
s. B. Donner 2nd, 32, *d*. by Old Jack.

SINGLETON—324.

Lambed 1883 ; *br*. Thomas Brown, *s*. No 2, 211,
d. by Broadfield 54.

SIR ARTHUR—325.

Lambed 1875 ; *br*. William Lane, *ow*. Robert Jacobs,
s. No. 20 (Garne's).

SIR CHARLES—326.

Lambed 1889 ; *br*. Charles Barton, *ow*. George Freeman.

SIR CHARLES—327.

Lambed 1889 ; *br*. C. Barton, *ow*. Wm. Houlton, *s*. No. 7 (Lane's).

SIR CHARLES—328.

Lambed 1890 ; *br*. C. Barton, *ow*. J. J. Godwin, *s*. Noble 196,
d. by Prince 282.

SIR JAMES—329.

Prize, 1860, 1st R. A. S., Canterbury.

Lambed 1859 ; *br.* James Walker, *ow.* Thomas Brown.

SIR JAMES—330.

Lambed about 1868 ; *br.* E. Handy, *ow.* J. Walker.

SIR JOHN—331.

Lambed 1886 ; *br.* Robert Garne, *ow.* John Waine, *s.* Timothy 377.

SIR ROBERT—332.

Lambed 1890 ; *br.* R. Jacobs, *ow.* W. J. Smith, *s.* Prior 285.

SIR STANLEY—333.

Lambed 1890 ; *br.* R. Jacobs, *ow.* W. J. Smith, *s.* Young Lane 399.

SIR THOMAS—334.

Lambed 1890 ; *br.* R. Jacobs, *ow.* W. J. Smith, *s.* Young Lane 399.

SIR WILLIAM—335.

Lambed 1868 ; *br.* William Lane, *ow.* Thomas Brown.

SIR WILLIAM—336.

Lambed 1883 ; *br.* Charles Barton, *s.* O. Fowler 258,
    d. by Moses 191.

SIR WILLIAM, Jun.—337.

Lambed 1871 ; *br.* Thomas Brown, *s.* Sir William 335,
    d. by Lord Lyon 181.

SIR WILLIAM II.—338.

Lambed 1885 ; *br.* Charles Barton, *s.* Sir William 336.

SLASHER—339.

Lambed 1885 ; *br.* Thomas Brown, *s.* Singleton 324.

SLAUGHTER—340.

Lambed 1887 ; *br.* Robert Garne, *s.* Rupert 309.

SLAUGHTER—341.

Lambed 1890 ; *br.* Edward Brain, *ow.* John P. Wakefield.

SON OF BANDY—342.

Lambed about 1855 ; *br.* W. Lane. *s.* Bandy.

SON OF No. 9—343.

Lambed 1885 ; *br.* Russell Swanwick, *s.* No. 9, 228.

SON OF GAY LAD—344.

Lambed 1884 ; *br.* Thomas Brown, *s.* Gay Lad 126,
d. by Example 102.

SON OF GREY NOBLEMAN—345.

Lambed 1879 ; *br.* Thomas Brown, *s.* Grey Nobleman 142.

SON OF K. B.—346.

Lambed 1881 ; *br.* Thomas Brown, *s.* K. B. 159.

SON OF No. 8—347.

Lambed 1881 ; *br.* Thomas Brown, *s.* No. 8, 225.

SPOT—348.

Lambed 1880 ; *br.* Thomas Brown, *s.* Marham Blackfoot 185.

SPOT—349.

Lambed 1885 ; *br.* William Lane, *ow.* Robert Jacobs.

STURDY—350.

Lambed 1884 ; *br.* Thomas Brown, *s.* Spot 348, *d.* by K. B. 159.

SUBSTANCE—351.

Lambed about 1855 ; *br.* E. Handy, *s.* Windsor 386.

SULTAN—352.

Lambed 1880 ; *br.* Thomas Brown, *s.* Marham Blackfoot, 185,
d. by Grey Lane 142.

SURPRISE—353.

Lambed 1883 ; *br.* Thomas Brown, *s.* Royal Kilburn 302,
d. by Ensign 99.

TANGLEY, Jun.—354.

Lambed 1888 ; *br.* Hugh Aylmer, *s.* Tangley.

TANGLEY II.—355.

Lambed 1888 ; *br.* Hugh Aylmer, *s.* Tangley.

TAYNTON—356.

Lambed 1885 ; *br.* Robert Jacobs, *s.* Royal Reading 305,
d. by Duke of Carlisle 93.

T. DONNER—357.

Lambed 1883 ; *br.* Charles Barton, *s.* Donner 91,
d. by Thick Neck.

T. DONNER II.—358.

Lambed 1888 ; *br.* Charles Barton, *s.* T. Donner 357,
d. by Old Jack.

THE BIG NECKED SHEEP—359.

Lambed about 1864; *br.* W. Hewer.

THE BRITISH WORKMAN—360.

Lambed 1889; *br.* Frederick Craddock, *s.* Old Jacob.

THE DOCTOR—361.

Lambed 1869; *br.* Thomas Brown, *s.* Lord Lyon 181.

THE DON—362.

Lambed 1889; *br.* Robert Garne, *s.* Ablington 4.

THE DUKE—363.

Lambed 1890; *br.* George Bagnall & Son, *s.* by Jumbo 155,
d. by Big Bill 44.

THE EASTINGTON DONNER—364.

Lambed 1891; *br.* Frederick Craddock, *s.* the 1890 Donner 369,
d. by Old Jacob.

THE FIFTY GUINEA RAM—365.

Lambed 1880; *br.* William Lane, *ow.* Russell Swanwick,
s. Stockwell.

THE LONG STRAIGHT SHEEP—366.

Lambed about 1870; *br.* W. Lane.

THE MARSHAL—367.

Lambed 1870; *br.* Thomas Brown, *s.* Sir William 335,
d. by Lord Lyon 181.

THE MARSHAL—368.

Lambed 1874 ; *br.* W. Lane, *ow.* Robert Garne,

s. The Long Straight Sheep 366,

d. by Cotswold King 80.

THE 1890 DONNER—369.

Lambed 1890 ; *br.* R. Garne, *ow.* F. Craddock, *s.* Donner 2nd, 92.

THE PRINCE—370.

Prize, 1869, 2nd R. A. S., Manchester.

Lambed 1868 ; *br.* Thomas Gillett, *ow.* Thomas Brown.

THE SQUIRE—371.

Lambed 1870 ; *br.* Robert Garne, *s.* Luckless 182.

THE SECOND SQUIRE—372.

Lambed 1873 ; *br.* Robert Garne, *s.* The Squire 371.

THICK JOHN—373.

Lambed 1886 ; *br.* Robert Jacobs, *ow.* M. Savidge,

s. Long John 179, *d.* by Royal Reading 305.

THICKSET—374.

Lambed 1876 ; *br.* Thomas Brown, *s.* No. 3, 2nd, 218.

THUNDER—375.

Lambed 1888 ; *br.* Russell Swanwick, *s.* Donner,

d. by the 50 Guinea Ram 365.

TICHBORNE—376.

Lambed 1881 ; *br.* Charles Barton, *s.* No. 4, 220.

TIMOTHY—377.

Lambed 1879 ; *br.* Robert Garne, *s.* No. 1, 203.

TIMOTHY—378.

Lambed 1890 ; *br.* Henry Clark, *ow.* C. E. Clark, *s.* Captain 64,

TINY—379.

Lambed 1870 ; *br.* Robert Garne, *s.* No Name 200.

VICAR OF BROADFIELD—380.

Lambed 1887 ; *br.* William Lane, *ow.* Walter Powell.

VICTOR—381.

Lambed 1891 ; *br.* O. F. Williams, *ow.* Mrs. Burrell.

WARHAM—382.

Lambed 1890 ; *br.* J. H. Yeomans, *ow.* E. Powell,
s. Fyfield 121, *d.* by Abbot 2.

WEST DEREHAM—383.

Lambed 1889 ; *br.* Hugh Aylmer, *ow.* Russell Swanwick,
s. Aldsworth.

WEST DEREHAM, 1891—384.

Lambed 1890 ; *br.* Hugh Aylmer, *ow.* Russell Swanwick,
s. Young Broadfield.

WHITE NOBLEMAN—385.

Lambed 1875 ; *br.* Thomas Brown,
s. Lane's Two-shear Sheep 167.

WINDSOR—386.

Lambed about 1854 ; *br.* Joseph Craddock,

s. The Windsor Shearling, 1851.

WINDSOR CASTLE—387.

Lambed 1889 ; *br.* Robert Garne, *ow.* John Garne,

s. Windsor II. 390.

WINDSOR RESERVE—388.

Lambed 1888 ; *br.* Russell Swanwick, *s.* Northampton I. 202.

WINDSOR RESERVE, TWO-SHEAR—389.

Lambed 1887 ; *br.* Russell Swanwick, *s.* No. 10, 231.

WINDSOR II.—390.

Prize, 1889, 2nd R. A. S., Windsor.

Lambed 1887 ; *br.* Robert Garne, *ow.* T. Brown,

s. Master Charles 186, *d.* Lady by Westwell.

WINDSOR III.—391.

Prize, 1889, 3rd R. A. S., Windsor.

Lambed 1887 ; *br.* Robert Garne, *s.* Excelsior 103,

d. Countess by No. 2, 209.

YOUNG ALDSWORTH—392.

Lambed 1888 ; *br.* Hugh Aylmer, *s.* Privateer.

YOUNG ALDSWORTH II.—393.

Lambed 1887 ; *br.* Hugh Aylmer, *s.* No. 6 (Garne's).

YOUNG CAPTAIN—394.

Lambed 1876 ; *br.* Robert Garne, *s.* W.

YOUNG CROP EAR—395.

Lambed 1890 ; *br.* H. J. Elwes, *s.* No. 10 (Garne's).

YOUNG DONNER—396.

Lambed 1887 ; *br.* Charles Barton, *ow.* William Houlton,
s. T. Donner 357, *d.* by Sir William 336.

YOUNG GREY—397.

Lambed 1889 ; *br.* Hugh Aylmer, *s.* Grey Face.

YOUNG JOE—398.

Lambed 1889 ; *br.* Henry Clark, *s.* Old Joe.

YOUNG LANE—399.

Lambed 1889 ; *br.* Robert Jacobs, *s.* Spot 349.
d. by Collegian 73.

YOUNG LYON—400.

Lambed 1875 ; *br.* Robert Garne, *s.* No 5, 222.

YOUNG PLYMOUTH—401.

Lambed 1890 ; *br.* George Freeman, *s.* Plymouth 274.

YOUNG PROFESSOR—402.

Lambed 1888 ; *br.* Robert Garne, *ow.* George Freeman.
s. Professor 288.

YOUNG SUBSTANCE—403.

Lambed about 1860 ; *br.* E. Handy, *s.* Substance 351.

REGISTER OF EWES.

In the Registry of Cotswold Ewes "*br.*" is used to signify "Breeder and Owner" where the animal has not changed ownership.

The other abbreviations are "*ow.*" for Owner; "*s.*" for Sire; "*d.*" for Dam; "*2nd d.*" for Grand-dam; "*3rd d.*" for Great-grand-dam; "*4th d.*" for Great-great-grand-dam.

Registered Cotswold Ewes are distinguished by a Flock Letter and Number, and by the Mark or Number in the Ear.

A 1, 300, lambed 1877 ; *br.* R. Garne, *s.* Experiment 105.

A 2, 30, lambed about 1881 ; *br.* R. Garne, *s.* No. 2, 209.

A 3, 90, lambed about 1883 ; *br.* W. Lane, *ow.* R. Garne,
s. Jumbo 154.

BF 1, 16, lambed 1881 ; *br.* C. Barton, *s.* No. 10, 230.

BF 2, 26, lambed 1883 ; *br.* C. Barton, *s.* Gramp 134.

BF 3, 33, lambed 1883 ; *br.* C. Barton, *s.* Prince 282.

BF 4, 41, lambed 1883 ; *br.* C. Barton, *s.* Donner 91.

BF 5, 15, lambed 1884 ; *br.* C. Barton, *s.* O. Fowler 258.

BF 6, 24, lambed 1884 ; *br.* C. Barton, *s.* No. 10, 230.

BF 7, 46, lambed 1884 ; *br.* C. Barton, *s.* O. Fowler 258.

BF 8, 50, lambed 1884 ; *br.* C. Barton, *s.* Tichborne 376.

BF 9, 45, lambed 1885 ; *br.* C. Barton, *s.* T. Donner 357.

BF 10, 52, lambed 1885 ; *br.* C. Barton, *s.* Tichborne 376.

BF 11, 53, lambed 1885 ; *br.* C. Barton, *s.* Jack 148.

BF 12, 55, lambed 1885 ; *br.* C. Barton, *s.* O. Fowler 258.

BF 13, 57, lambed 1885 ; *br.* C. Barton, *s.* O. Fowler 258.

BF 14, 58, lambed 1885 ; *br.* C. Barton, *s.* Jack 148.

BF 15, 60, lambed 1885 ; *br.* C. Barton, *s.* Jack 148.

BF 16, 20, lambed 1886 ; *br.* C. Barton, *s.* O. Fowler 258.

BF 17, 27, lambed 1886 ; *br.* C. Barton, *s.* B. Donner 31.

BF 18, 29, lambed 1886 ; *br.* C. Barton, *s.* Prince 282.

BF 19, 34, lambed 1886 ; *br.* C. Barton, *s.* No. 1, 206.

BF 20, 35, lambed 1886 ; *br.* C. Barton, *s.* No. 1, 206.

BF 21, 37, lambed 1886 ; *br.* C. Barton, *s.* Jack 148.

BF 22, 39, lambed 1886 ; *br.* C. Barton, *s.* B. Donner 31.

BF 23, 14, lambed 1887 ; *br.* C. Barton, *s.* B. Donner 31.

BF 24, 18, lambed 1887 ; *br.* C Barton, *s.* King 161.

BF 25, 25, lambed 1887 ; *br.* C. Barton, *s.* B. Donner 31.

BF 26, 31, lambed 1887 ; *br.* C. Barton, *s.* B. Donner 31.

BF 27, 32, lambed 1887 ; *br.* C. Barton, *s.* Bill 45.

BF 28, 36, lambed 1887 ; *br.* C. Barton, *s.* No. 1, 206.

BF 29, 42, lambed 1887 ; *br.* C. Barton, *s.* King 161.

BF 30, 44, lambed 1887 ; *br.* C. Barton, *s.* B. Donner 31.

BF 31, 47, lambed 1887 ; *br.* C. Barton, *s.* No. 1, 206.

BF 32, 48, lambed 1887 ; *br.* C. Barton, *s.* No. 1, 206.

BF 33, 49, lambed 1887 ; *br.* C. Barton, *s.* King 161.

BF 34, 63, lambed 1887 ; *br.* C. Barton, *s.* Sir William 336.

BF 35, 17, lambed 1888 ; *br.* C. Barton, *s.* Bill 45.

BF 36, 19, lambed 1888 ; *br.* C. Barton, *s.* Broadfield 56.

BF 37, 30, lambed 1888 ; *br.* C. Barton, *s.* B. Donner 31.

BF 38, 43, lambed 1888 ; *br.* C. Barton, *s.* No. 30, 251.

BF 39, 61, lambed 1888 ; *br.* C. Barton, *s.* Broadfield 56.

BF 40, 62, lambed 1888 ; *br.* C. Barton, *s.* Broadfield 56.

BF 41, 21, lambed 1889 ; *br.* C. Barton, *s.* Bill 45.

BF 42, 22, lambed 1889 ; *br.* C. Barton, *s.* Noble 196.

BF 43, 23, lambed 1889 ; *br.* C. Barton, *s.* B. Donner II. 32.

BF 44, 28, lambed 1889 ; *br.* C. Barton, *s.* Bill 45.

BF 45, 38, lambed 1889 ; *br.* C. Barton, *s.* Noble 196.

BF 46, 40, lambed 1889 ; *br.* C. Barton, *s.* Noble 196.

BF 47, 54, lambed 1889 ; *br.* C. Barton, *s.* Captain 64.

BF 48, 56, lambed 1889 ; *br.* C. Barton, *s.* Broadfield 56.

BF 49, 59, lambed 1889 ; *br.* C. Barton, *s.* Broadfield 56.

BF 50, 51, lambed 1889 ; *br.* C. Barton, *s.* Gramp II. 135.

M 1, 1, lambed 1883 ; *br.* T. Brown, *s.* Royal Derby, 300,
              d. by Lane's Old Sheep 166.

M 2, 2, lambed 1883 ; *br.* T. Brown, *s.* Royal Derby 300,
              d. by K. B. 159.

M 4, 4, lambed 1883 ; *br.* T. Brown, *s.* Gay Lad 126,
              d. by Boreas 51.

M 5, 5, lambed 1883 ; *br.* T. Brown, *s.* No. 20, 246,
              d. by Ensign 99.

M 6, 6, lambed 1883 ; *br.* T. Brown, *s.* Gay Lad 126,
              d. by No. 42, 256.

M 12, 12, lambed 1883 ; *br.* T. Brown, *s.* Spot 348,
              d. by No. 4, 219.

M 13, 13, lambed 1883 ; *br.* T. Brown, *s.* Reserve 294,
              d. by Royal Kilburn 302.

M 16, 16, lambed 1883 ; *br.* T. Brown, *s.* Gay Lad 126,
              d. by No No 201.

M 17, 17, lambed 1883 ; *br.* T. Brown, *s.* Commodore 76,
              d. by Blue Lane 48.

M 18, 18, lambed 1883 ; *br.* T. Brown, *s.* Spot 348,
              d. by Grey Lane 141.

M 19, 19, lambed 1883 ; *br.* T. Brown, *s.* Spot 348,
              d. by Grey Lane 141.

M 20, 20, lambed 1883 ; *br.* T. Brown, *s.* Reserve 294,
              d. by Prince 281.

M 27, 27, lambed 1883 ; *br.* T. Brown, *s.* No. 20, 246,

    d. by No. 3, 2nd, 218.

M 29, 29, lambed 1883 ; *br.* T. Brown, *s.* Gay Lad 126,

    d. General 129.

M 32, 32, lambed 1883 ; *br.* T. Brown, *s.* Gay Lad 126,

    d. by Boreas 51.

M 41, 41, lambed 1883 ; *br.* T. Brown, *s.* Reserve 294,

    d. by Blue Lane 48.

M 82, 82, lambed 1883 ; *br.* T. Brown, *s.* No. 9, 226,

    d. Old Grey, by No. 4, 219.

M 87, 87, lambed 1883 ; *br.* T. Brown, *s.* Royal Derby 300,

    d. by Broadfield 54.

M 88, 88, lambed 1883 ; *br.* T. Brown, *s.* Commodore 76,

    d. by Marham Blackfoot 185.

M 89, 89, lambed 1883 ; *br.* T. Brown, *s.* No. 9, 226,

    d. by Royal Kilburn, 302.

M 90, 90, lambed 1883 ; *br.* T. Brown, *s.* Royal Kilburn 302,

    d. by White Nobleman, 385.

M 92, 92, lambed 1883 ; *br.* T. Brown, *s.* Royal Derby 300,

    d. by Marham Blackfoot 185.

M 94, 94, lambed 1883 ; *br.* T. Brown, *s.* Commodore 76,

    d. by Broadfield 54.

M 102, 102, lambed 1883 ; *br.* T. Brown, *s.* Quality 289,

    d. by Marham Blackfoot 185.

M 106, 106, lambed 1883 ; *br.* T. Brown, *s.* Quality 289,
d. by Reserve 294.

M 114, 114, lambed 1883 ; *br.* T. Brown, *s.* Spot 348,
d. by Broadfield 54.

M 118, 118, lambed 1883 ; *br.* T. Brown, *s.* Reserve 294,
d. by No. 4, 219.

M 121, 121, lambed 1883 ; *br.* T. Brown, *s.* Spot 348,
d. by Reserve 294.

M 126, 126, lambed 1883 ; *br.* T. Brown, *s.* Royal Derby 300,
d. by General 129.

M 128, 128, lambed 1883 ; *br.* T. Brown, *s.* Gay Lad 126,
d. by Grey Lane 141.

M 130, 130, lambed 1883 ; *br.* T. Brown, *s.* Reserve 294,
d. by Broadfield 54.

M 131, 131, lambed 1883 ; *br.* T. Brown, *s.* Spot 348,
d. by Reserve 294.

M 134, 134, lambed 1883 ; *br.* T. Brown, *s.* Spot 348,
d. by No. 2, 210.

M 142, 142, lambed 1883 ; *br.* T. Brown, *s.* Quality 289,
d. by Grey Nobleman 142.

M 145, 145, lambed 1883 ; *br.* T. Brown, *s.* Commodore 76,
d. by No. 42, 256

M 147, 147, lambed 1883 ; *br.* T. Brown, *s.* Quality 289,
d. by Nobleman 199.

M 148, 148, lambed 1883 ; *br.* T. Brown, *s.* Commodore 76,
d. by Broadfield 54.

M 150, 150, lambed 1883 ; *br.* T. Brown, *s.* Gay Lad 126,
d. by Reserve 294.

M 156, 156, lambed 1883 ; *br.* T. Brown, *s.* Spot 348,
d. by Marham Blackfoot 185.

M 163, 163, lambed 1883 ; *br.* T. Brown, *s.* Quality 289,
d. by Admiral 6.

M 164, 164, lambed 1883 ; *br.* T. Brown, *s.* Royal Derby 300,
d. by Broadfield 54.

M 166, 166, lambed 1883 ; *br.* T. Brown, *s.* Commodore 76,
d. by No. 8, 225.

M 172, 172, lambed 1883 ; *br.* T. Brown, *s.* Commodore 76,
d. by Grey Lane 141.

M 173, 173, lambed 1883 ; *br.* T. Brown, *s.* Son of K. B. 346,
d. by Morton 190.

M 174, 174, lambed 1883 ; *br.* T. Brown, *s.* Experience 106,
d. by Grey Nobleman 142.

M 182. 182, lambed 1883 ; *br.* T. Brown, *s.* Reserve 294,
d. by Royal Kilburn 302.

M 184, 184, lambed 1883 ; *br.* T. Brown, *s.* Commodore 76,
d. by Little Doctor 177.

M 199, 199, lambed 1883 ; *br.* T. Brown, *s.* Commodore 76,
d. by No. 42, 256.

M 200, 200, lambed 1883 ; *br.* T. Brown, *s.* Quality 289,
     d. by Grey Nobleman 142.

M 206, 206, lambed 1883 ; *br.* T. Brown, *s* Royal Kilburn 302,
     d. Broadfield 54.

M 208, 208, lambed 1883 ; *br.* T. Brown, *s.* Reserve 294,
     d. by Grey Nobleman 142.

M 209, 209, lambed 1883 ; *br.* T. Brown, *s.* Spot 348,
     d. by No. 2, 210.

M 216, 216, lambed 1883 ; *br.* T. Brown, *s.* Son of No. 8, 347,
     d. by General 129.

M 233, 233, lambed 1884 ; *br.* T. Brown, *s.* Beacon 34,
     d. by No. 42, 256.

M 238, 238, lambed 1884 ; *br.* T. Brown, *s.* Example 102,
     d. by Lane's Old Sheep 166.

M 241, 241, lambed 1884 ; *br.* T. Brown, *s.* Haylock 146,
     d. by Royal Kilburn 302.

M 244, 244, lambed 1884 ; *br.* T. Brown, *s.* Spot 348,
     d. by Reserve 294.

M 245, 245, lambed 1884 ; *br.* T. Brown, *s.* Spot 348,
     d. by Royal Kilburn 302.

M 247, 247, lambed 1884 ; *br.* T. Brown, *s.* Spot 348,
     d. by Experience 106.

M 251, 251, lambed 1884 ; *br.* T. Brown, *s.* Duke of York 94,
     d. by Broadfield 54.

M 252, 252, lambed 1884 ; *br.* T. Brown, *s.* Example 102,
d. by No. 8, 225.

M 254, 254, lambed 1884 ; *br* T. Brown, *s.* Beacon 34,
d. by Broadfield 54.

M 256, 256, lambed 1884 ; *br.* T. Brown, *s.* Duke of York 94,
d. by Broadfield 54.

M 258, 258, lambed 1884 ; *br.* T. Brown, *s.* Beacon 34,
d. by Blue Lane 48.

M 268, 268, lambed 1884 ; *br.* T. Brown, *s.* Example 102,
d. by Blue Lane 48.

M 272, 272, lambed 1884 ; *br.* T. Brown, *s.* Kent 163.

M 275, 275, lambed 1884 ; *br.* T. Brown, *s.* Duke of York 94,
d. by No. 1, 205.

M 279, 279, lambed 1884 ; *br.* T. Brown, *s.* Experience 106,
d. by Grey Nobleman 142.

M 280, 280, lambed 1884 ; *br.* T. Brown, *s.* Gay Lad 126,
d. by Dick Allen 87.

M 281, 281, lambed 1884 ; *br.* T. Brown, *s.* Beacon 34,
d. by No. 8, 225.

M 284, 284, lambed 1884 ; *br.* T. Brown, *s.* Spot 348,
d. by Broadfield 54.

M 288, 288, lambed 1884 ; *br.* T. Brown, *s.* Commodore 76,
d. by Broadfield 54.

M 289, 289, lambed 1884 ; *br.* T. Brown, *s.* Gay Lad 126,
d. by Reserve 294.

M 301, 301, lambed 1884 ; *br.* T. Brown, *s.* Experience 106,
d. by Reserve 294.

M 302, 302, lambed 1884 ; *br.* T. Brown, *s.* Kent 163,
d. by Nobleman 199.

M 304, 304, lambed 1884 ; *br.* T. Brown, *s.* Spot 348,
d. by Broadback 53.

M 305, 305, lambed 1884 ; *br.* T. Brown, *s.* Spot 348,
d. by Reserve 294.

M 310, 310, lambed 1884 ; *br.* T. Brown, *s.* Experience 106,
d. by Garne's Blackfoot 123.

M 312, 312, lambed 1884 ; *br.* T. Brown, *s.* Beacon 34,
d. by Royal Kilburn 302.

M 313, 313, lambed 1884 ; *br.* T. Brown, *s.* Commodore 76,
d. by Broadfield 54.

M 315, 315, lambed 1884 ; *br.* T. Brown, *s.* Kent 163,
d. by No. 42, 256.

M 319, 319, lambed 1884 ; *br.* T. Brown, *s.* Spot 348,
d. by Nobleman 199.

M 322, 322, lambed 1884 ; *br.* T. Brown, *s.* Haylock 146,
d. by Sultan 352.

M 327, 327, lambed 1884 ; *br.* T. Brown, *s.* Royal Kilburn 302,
d. by Broadfield 54.

M 334, 334, lambed 1884 ; *br.* T. Brown, *s.* Beacon 34,
     d. by No. 1, 205.

M 339, 339, lambed 1884 ; *br.* T. Brown, *s.* Commodore 76,
     d. by Broadfield 54.

M 340, 340, lambed 1884 ; *br.* T. Brown, *s.* Haylock 146,
     d. by Lane's Old Sheep, 166.

M 342, 342, lambed 1884 ; *br.* T. Brown, *s.* Beacon 34,
     d. by Royal Kilburn 302.

M 348, 348, lambed 1884 ; *br.* T. Brown, *s.* Example 102,
     d. by Broadfield 54.

M 351, 351, lambed 1884 ; *br.* T. Brown, *s.* Royal Kilburn 302,
     d. by Reserve 294.

M 358, 358, lambed 1884 ; *br.* T. Brown, *s.* Experience 106,
     d. by Broadback 53.

M 364, 364, lambed 1884 ; *br* T. Brown, *s.* Duke of York 94.

M 365, 365, lambed 1884 ; *br.* T. Brown, *s.* Duke of York 94.

M 366, 366, lambed 1884 ; *br.* T. Brown, *s.* Gay Lad 126,
     d. by Son of Grey Nobleman 345.

M 370, 370, lambed 1884 ; *br.* T. Brown, *s.* Royal Kilburn 302,
     d. by Grey Son of Little Doctor 144.

M 372, 372, lambed 1884 ; *br.* T. Brown, *s.* Duke of York 94,
     d. by Admiral 6.

M 388, 388, lambed 1884 ; *br.* T. Brown, *s.* Spot 348,
     d. by Broadback 53.

M 392, 392, lambed 1884; *br.* T. Brown, *s.* Example 102, *d.* by Reserve 294.

M 397, 397, lambed 1884; *br.* T. Brown, *s.* Duke of York 94, *d.* by K. B. 159.

M 399, 399, lambed 1884; *br.* T Brown, *s.* Haylock 146, *d.* by Admiral 6.

M 406, 406, lambed 1884; *br.* T. Brown, *s.* No. 20, 246, *d.* by Royal Kilburn 302.

M 413, 413, lambed 1884; *br.* T. Brown, *s.* Commodore 76, *d.* by Grey Son of Broadfield 143.

M 423, 423, lambed 1884; *br.* T. Brown, *s.* No. 20, 246, *d.* by General 129.

M 428, 428, lambed 1884; *br.* T. Brown, *s.* No. 20, 246, *d.* by Broadfield 54.

M 429, 429, lambed 1884; *br.* T. Brown, *s.* No. 20, 246, *d.* by Royal Kilburn, 302.

M 430, 430, lambed 1884; *br.* T. Brown, *s.* No 20, 246, *d.* by Grey Lane 142.

M 435, 435, lambed 1884; *br.* T. Brown, *s.* No. 20, 246, *d.* by No. 1, 204.

M 451, 451, lambed 1885; *br.* T. Brown, *s.* Singleton 324, *d.* by Marham Blackfoot 185.

M 454, 454, lambed 1885; *br.* T. Brown, *s.* Kent 163, *d.* by Grey Nobleman 142

M 465, 465, lambed 1885 ; *br.* T. Brown, *s.* Lion 175,
          d. by No. 42, 256.

M 472, 472, lambed 1885 ; *br.* T. Brown, *s.* Kent 163,
          d. by Broadfield 54.

M 473, 473, lambed 1885 ; *br.* T. Brown. *s.* Lion 175,
          d. by Reserve 294.

M 478, 478, lambed 1885 ; *br.* T. Brown, *s.* Spot 348,
          d. by Reserve 294.

M 479, 479, lambed 1885 ; *br.* T. Brown, *s.* Surprise 353,
          d. by Reserve 294.

M 480, 480, lambed 1885 ; *br.* T. Brown, *s.* Rear-Admiral 292,
          d. by Marham Blackfoot 185.

M 481, 481, lambed 1885 ; *br.* T. Brown, *s.* Kent 163,
          d. by Marham Blackfoot, 185.

M 485, 485, lambed 1885 ; *br.* T. Brown, *s.* Lion 175,
          d. by No. 42, 256.

M 496, 496, lambed 1885 ; *br.* T. Brown, *s.* Example 102,
          d. by Reserve 294.

M 497, 497, lambed 1885 ; *br.* T. Brown, *s.* Kent 163,
          d. by Grey Son of Broadfield, 143.

M 498, 498, lambed 1885 ; *br.* T. Brown, *s.* Kent 163,
          d. by Marham Blackfoot 185.

M 504, 504, lambed 1885 ; *br.* T. Brown, *s.* Kent 163,
          d. by Marham Blackfoot 185.

M 506, 506, lambed 1885 ; *br.* T. Brown, *s.* Kent 163,
d. by Sultan 352.

M 507, 507, lambed 1885 ; *br.* T. Brown, *s.* Lion 175,
d. by Admiral 6.

M 508, 508, lambed 1885 ; *br.* T. Brown, *s.* Lion 175,
d. by No. 2, 210.

M 513, 513, lambed 1885 ; *br.* T. Brown, *s.* Example 102,
d. by Royal Kilburn 302.

M 516, 516, lambed 1885 ; *br.* T. Brown, *s.* Spot 348,
d. by Broadfield 54.

M 519, 519, lambed 1885 ; *br.* T. Brown, *s.* Example 102,
d. by No. 2, 210.

M 520, 520, lambed 1885 ; *br.* T. Brown, *s.* Example 102,
d. by No. 2, 210.

M 523, 523, lambed 1885 ; *br.* T. Browp, *s.* Kent 163,
d. by No. 1, 204.

M 532, 532, lambed 1885 ; *br.* T. Brown, *s.* Lion 175,
d. by Marham Blackfoot 185.

M 533, 533, lambed 1885 ; *br.* T. Brown, *s.* Example 102,
d. by Sultan 352.

M 546, 546, lambed 1885 ; *br.* T. Brown, *s.* Rear-Admiral 292,
d. by Reserve 294.

M 555, 525, lambed 1885 ; *br.* T. Brown, *s.* Singleton 324,
d. by Broadfield 54.

M 562, 562, lambed 1885 ; *br.* T. Brown, *s.* Beacon 34,
                     d. by Marham Blackfoot 185.

M 569, 569, lambed 1885 ; *br.* T. Brown, *s.* Kent 163,
                     d. by White Nobleman 385.

M 572, 572, lambed 1885 ; *br.* T. Brown, *s.* Surprise 353,
                     d. by Royal Kilburn 302.

M 581, 581, lambed 1885 ; *br.* T. Brown, *s.* Beacon 34,
                     d. by Royal Kilburn 302.

M 585, 585, lambed 1885 ; *br.* T. Brown, *s.* Example 102,
                     d. by Royal Kilburn 302.

M 587, 587, lambed 1885 ; *br.* T. Brown, *s.* Surprise 353,
                     d. by Grey Nobleman 142.

M 598, 598, lambed 1885 ; *br.* T. Brown, *s.* Example 102,
                     d. by Garne's Blackfoot 123.

M 599, 599, lambed 1885 ; *br.* T. Brown, *s.* Lion 175,
                     d. by Dick Allen 87.

M 600, 600, lambed 1885 ; *br.* T. Brown, *s.* Singleton 324,
                     d. by No. 8, 225.

M 601, 601, lambed 1885 ; *br.* T. Brown, *s.* Commodore 76,
                     d. by Reserve 294.

M 607, 607, lambed 1885 ; *br.* T. Brown, *s.* Singleton 324,
                     d. by Royal Kilburn 302.

M 608, 608, lambed 1885 ; *br.* T. Brown, *s.* Singleton 324,
                     d. by Royal Kilburn 302.

M 613, 613, lambed 1885 ; *br.* T. Brown, *s.* Example 102,
d. by Reserve 294.

M 614, 614, lambed 1885 ; *br.* T. Brown, *s.* Beacon 34,
d. by Royal Kilburn 302.

M 623, 623, lambed 1885 ; *br.* T. Brown, *s.* No. 2, 211,
d. M 94 by Commodore 76,
2nd d. by Broadfield 54.

M 625, 625, lambed 1885 ; *br.* T. Brown, *s.* Spot 348,
d. by General 129.

M 628, 628, lambed 1885 ; *br.* T. Brown, *s.* No. 2, 211,
d. M 92 by Royal Derby 300,
2nd d. by Marham Blackfoot 185.

M 633, 633, lambed 1885 ; *br.* T. Brown, *s.* No 2, 211,
d. M 156 by Spot 348,
2nd d. by Marham Blackfoot 185.

M 636, 636, lambed 1885 ; *br.* T. Brown, *s.* Beacon 34,
d. M 121 by Spot 348,
2nd d. by Reserve 294.

M 638, 638, lambed 1885 ; *br.* T. Brown, *s.* Lion 175,
d. by Broadfield 54.

M 649, 649, lambed 1885 ; *br.* T. Brown, *s.* Surprise 353,
d. by Broadfield 54.

M 657, 657, lambed 1885 ; *br.* T. Brown, *s.* Commodore 76,
d. by Reserve 294.

M 660, 660, lambed 1885 ; *br.* T. Brown, *s.* Beacon 34.

M 663, 663, lambed 1885 ; *br.* T. Brown, *s.* Surprise 353.

M 673, 673, lambed 1885 ; *br.* T. Brown, *s.* Example 102,
      d. M 164 by Royal Derby 300,
      2nd d. by Broadfield 54.

M 674, 674, lambed 1885 ; *br.* T. Brown, *s.* No. 2, 211,
      d. M 102 by Quality 289,
      2nd d. by Marham Blackfoot 185.

M 684, 684, lambed 1885 ; *br.* T. Brown, *s.* Beacon 34,
      d. by Admiral 6.

M 685, 685, lambed 1885 ; *br.* T. Brown, *s.* Beacon 34,
      d. by Admiral 6.

M 704, 704, lambed 1885 ; *br.* T. Brown, *s.* Surprise 353,
      d. by Experience 106.

M 705, 705, lambed 1885 ; *br.* T. Brown, *s.* No. 2, 211,
      d. M 106 by Quality 289,
      2nd d. by Reserve 294.

M 709, 709, lambed 1885 ; *br.* T. Brown, *s.* Commodore 76,
      d. by Royal Kilburn 302.

M 723, 723, lambed 1885 ; *br.* T. Brown, *s.* Example 102,
      d. M 182 by Reserve 294,
      2nd d. by Royal Kilburn 302.

M 752, 752, lambed 1885 ; *br.* T. Brown, *s.* No. 20, 246,
      d. by White Nobleman 385.

M 759, 759, lambed 1886 ; *br.* T. Brown, *s.* Sturdy 350,
      d. by No. 8, 225.

M 760, 760, lambed 1886 ; *br.* T. Brown, *s.* Garnet 124,
d. by Royal Kilburn 302.

M 763, 763, lambed 1886 ; *br.* T. Brown, *s.* Paragon, 268,
d. M 17 by Commodore 76,
2nd d. by Blue Lane 48.

M 764, 764, lambed 1886 ; *br.* T. Brown, *s.* Garnet 124,
d. by Admiral 6.

M 771, 771, lambed 1886 ; *br.* T. Brown, *s.* Excelsior 103,
d. M 5 by No. 20, 246,
2nd d. by Ensign 99.

M 773, 773, lambed 1886 ; *br.* T. Brown, *s.* Ronald 297,
d. M 305 by Spot 348,
2nd d. by Reserve 294.

M 776, 776, lambed 1886 ; *br.* T. Brown, *s.* Excelsior 103,
d. by No. 42, 256.

M 779, 779, lambed 1886 ; *br.* T. Brown, *s.* Sturdy 350,
d. by Admiral 6.

M 783, 783, lambed 1886 ; *br.* T. Brown, *s.* Spot 348,
d. by Experience 106.

M 786, 786, lambed 1886 ; *br.* T. Brown, *s.* Garnet 124,
d. by Admiral 6.

M 788, 788, lambed 1886 ; *br.* T. Brown, *s.* Favorite 109,
d by Dick Allen 87.

M 789, 789, lambed 1886 ; *br.* T. Brown, *s.* Garnet 124.

M 793, 793, lambed 1886 ; *br.* T. Brown, *s.* Favorite 109,
d. by Example 102.

M 794, 794, lambed 1886 ; *br.* T. Brown, *s.* Ronald 297,
d. by Sultan 352.

M 796, 796, lambed 1886 ; *br.* T. Brown, *s.* Ronald 297,
d. by Admiral 6.

M 797, 797, lambed 1886 ; *br.* T. Brown, *s.* Ronald 297,
d. by Example 102.

M 798, 798, lambed 1886 ; *br.* T. Brown, *s.* Ronald 297,
d. by Example 102.

M 802, 802, lambed 1886 ; *br.* T. Brown, *s.* Ronald 297,
d. by Broadback 53.

M 803, 803, lambed 1886 ; *br.* T. Brown, *s.* Favorite 109,
d. by Reserve 294.

M 804, 804, lambed 1886 ; *br.* T. Brown, *s.* Sturdy 350,
d. by Admiral 6.

M 808, 808, lambed 1886 ; *br.* T. Brown, *s.* Garnet 124,
d. by K. B. 159.

M 815, 815, lambed 1886 ; *br.* T. Brown, *s.* Paragon 268,
d. by Broadfield 54.

M 817, 870, lambed 1886 ; *br.* T. Brown, *s.* Sturdy 350,
d. by No. 2, 210.

M 821, 821, lambed 1886 ; *br.* T. Brown, *s.* Favorite 109,
d. by Royal Kilburn 302.

M 823, 823, lambed 1886 ; *br.* T. Brown, *s.* Sturdy 350,
           d. by Garne's Blackfoot 123.

M 824, 824, lambed 1886 ; *br.* T. Brown, *s.* Excelsior 103,
           d. by Broadback 53.

M 827, 827, lambed 1886 ; *br.* T. Brown, *s.* Favorite 109,
           d. M 156 by Spot 348,
           2nd d. by Marham Blackfoot 185.

M 830, 830, lambed 1886 ; *br.* T. Brown, *s.* Ronald 297,
           d. by Broadback 53.

M 831, 831, lambed 1886 ; *br.* T. Brown, *s.* Garnet 124,
           d. M 142 by Quality 289,
           2nd d. by Grey Nobleman 142.

M 832, 832, lambed 1886 ; *br.* T. Brown, *s.* Garnet 124,
           d. M 142 by Quality 289,
           2nd d. by Grey Nobleman 142.

M 836, 836, lambed 1886 ; *br.* T. Brown, *s.* Paragon 268,
           d. M 173 by Son of K. B. 346,
           2nd d. by Morton 190.

M 837, 837, lambed 1886 ; *br.* T. Brown, *s.* Spot 348,
           d. by Admiral 6.

M 838, 838, lambed 1886 ; *br.* T. Brown, *s.* Spot 348,
           d. by Admiral 6.

M 840, 840, lambed 1886 ; *br.* T. Brown, *s.* Spot 348,
           d. by Reserve 294.

M 843, 843, lambed 1886 ; *br.* T. Brown, *s.* Paragon 268,
           d. by Marham Blackfoot 185.

M 845, 845, lambed 1886 ; *br.* T. Brown, *s.* Sturdy 350,
    d. M 27 by No. 20, 246,
    2nd d. by No. 3, 2nd, 218.

M 851, 851, lambed 1886 ; *br.* T. Brown, *s.* Paragon 268,
    d. by No. 2, 210.

M 852, 852, lambed 1886 ; *br.* T. Brown, *s.* Paragon 268.
    d. by No. 2, 210.

M 857, 857, lambed 1886 ; *br.* T. Brown, *s.* Ronald 297,
    d. by Experience 106.

M 859, 859, lambed 1886 ; *br.* T. Brown, *s.* Favorite 109,
    d. by Broadfield 54.

M 860, 860, lambed 1886 ; *br.* T. Brown, *s.* Excelsior 103,
    d. by Admiral 6.

M 861, 861, lambed 1886 ; *br.* T. Brown, *s.* Favorite 109,
    d. M 41 by Reserve 294,
    2nd d. by Blue Lane 48.

M 862, 862, lambed 1886 ; *br.* T. Brown, *s.* Excelsior 103,
    d. by Admiral 6.

M 864, 864, lambed 1886 ; *br.* T. Brown, *s.* Spot 348,
    d. by Reserve 294.

M 865, 865, lambed 1886 ; *br.* T. Brown, *s.* Spot 348,
    d. by Reserve 294.

M 870, 817, lambed 1886 ; *br.* T. Brown, *s.* Excelsior 103,
    d. by General 129.

M 873, 873, lambed 1886 ; *br.* T. Brown, *s.* Spot 348,
    d. by K. B. 159.

M 876, 876, lambed 1886 ; *br.* T. Brown, *s.* Garnet 124,
d. by Broadfield 54.

M 878, 878, lambed 1886 ; *br.* T. Brown, *s.* by Favorite 109,
d. by Reserve 294.

M 887, 887, lambed 1886; *br.* T. Brown, *s.* Garnet 124,
d. M 118 by Reserve 294,
2nd d. by No. 4, 219.

M 889, 889, lambed 1886 ; *br.* T. Brown, *s.* Sturdy 350.
d. M 435 by No. 20, 246,
2nd d. by No. 1, 204.

M 897, 897, lambed 1886; *br.* T. Brown ; *s.* Sturdy 350,
d. by Broadfield 54.

M 903, 903, lambed 1886 ; *br.* T. Brown, *s.* Spot 348,
d. M 4 by Gay Lad 126,
2nd d. by Boreas 51.

M 913, 913, lambed 1886 ; *br.* T. Brown, *s.* Garnet 124,
d. M 126 by Royal Derby 300,
2nd d. by General 129.

M 920, 920, lambed 1886 ; *br.* T. Brown, *s.* Paragon 268,
d. M 216 by Son of No. 8, 347.
2nd d. by General 129.

M 925, 925, lambed 1886 ; *br.* T. Brown, *s.* Sturdy 350,
d. by Experience 106.

M 927, 927, lambed 1886; *br.* T. Brown, *s.* Garnet 124,
d. M 164 by Royal Derby 300,
2nd d. by Broadfield 54.

M 929, 929, lambed 1886 ; *br.* T. Brown, *s.* Paragon 268,
    d. by Reserve 294.

M 938, 938, lambed 1886 ; *br.* T. Brown, *s.* Favorite 109,
    d. by Example 102,
    2nd d. Old Grey by No. 4, 219.

M 940, 940, lambed 1886 ; *br.* T. Brown, *s.* Favorite 109,
    d. M 89 by No. 9, 226,
    2nd d. by Royal Kilburn 302.

M 946, 946, lambed 1886 ; *br.* T. Brown, *s.* Garnet 124,
    d. by Grey Nobleman 142.

M 951, 951, lambed 1886 ; *br.* T. Brown, *s.* Garnet 124,
    d. by Broadfield 54.

M 952, 952, lambed 1886 ; *br.* T. Brown, *s.* Garnet 124,
    d. by Broadfield 54.

M 958, 958, lambed 1886 ; *br.* T. Brown, *s.* Garnet 124,
    d. by Royal Kilburn 302.

M 964, 964, lambed 1886 ; *br.* T. Brown, *s.* Garnet 124,
    d. M 134 by Spot 348,
    2nd d. by No. 2, 210.

M 967, 967, lambed 1886 ; *br.* T. Brown, *s.* Paragon 268,
    d. by Royal Kilburn 302.

M 973, 973, lambed 1886 ; *br.* T. Brown, *s.* Garnet 124,
    d. by Reserve 294.

M 976, 976, lambed 1886 ; *br.* T. Brown, *s.* Paragon 268,
    d. M 172 by Commodore 76,
    2nd d. by Grey Lane 141.

M 989, 989, lambed 1886 ; *br.* T. Brown, *s.* Favorite 109,

d. M 163 by Quality 289,

2nd d. by Admiral 6.

M 996, 996, lambed 1886 ; *br.* T. Brown, *s.* Paragon 268,

d. by Broadfield 54.

M 1006, 1006, lambed 1886 ; *br.* T. Brown, *s.* Garnet 124,

d. M 315 by Kent 163,

2nd d by No. 42, 256.

M 1045, 1045, lambed 1887 ; *br.* T. Brown & Son, *s.* Slasher 339,

d. by Sultan 352.

M 1047, 1047, lambed 1887 ; *br.* T. Brown & Son, *s.* Cavalier 66,

d. M 312 by Beacon 54,

2nd d. by Royal Kilburn 302.

M 1054, 1054, lambed 1887 ; *br.* T. Brown & Son, *s.* Cavalier 66,

d. by Sultan 352.

M 1056, 1056, lambed 1887 ; *br.* T. Brown & Son, *s.* Slasher 339,

d. M 142 by Quality 289,

2nd d. by Grey Nobleman 142.

M 1058, 1058, lambed 1887 ; *br.* T. Brown & Son, *s.* Paragon 268,

d. M 288 by Commodore 76,

2nd d. by Broadfield 54.

M 1060, 1060, lambed 1887 ; *br.* T. Brown & Son, *s,* Spot 348,

d. by Reserve 294.

M 1067, 1067, lambed 1887 ; *br.* T. Brown & Son, *s.* Percy 270,

d. M 340 by Haylock 146,

2nd d. by Lane's Old Sheep 166.

M 1071, 1071, lambed 1887 ; *br.* T. Brown & Son, *s.* Citadel 70,
   d. by Admiral 6.

M 1078, 1078, lambed 1887 ; *br.* T. Brown & Son, *s.* Cavalier 66,
   d. M 150 by Gay Lad 126,
   2nd d. by Reserve 294.

M 1079, 1079, lambed 1887 ; *br.* T. Brown & Son, *s.* Paragon 268,
   d. M 27 by No. 20, 246,
   2nd d. by No. 3, 2nd, 218.

M 1085, 1085, lambed 1887 ; *br.* T. Brown & Son, *s.* Bayard 28,
   d. M 358 by Experience 106,
   2nd d. by Broadback 53.

M 1089, 1089, lambed 1887 ; *br.* T. Brown & Son, *s.* Slasher 339,
   d. M 392 by Example 102,
   2nd d. by Reserve 294.

M 1091, 1091, lambed 1887 ; *br.* T. Brown & Son, *s.* Paragon 268,
   d. by Reserve 294.

M 1093, 1093, lambed 1887 ; *br.* T. Brown & Son, *s.* Spot 348,
   d. M 147 by Quality 289,
   2nd d. by Nobleman 199.

M 1095, 1095, lambed 1887 ; *br.* T. Brown & Son, *s.* Shylock 315,
   d. M 2 by Royal Derby 300,
   2nd d. by K. B. 159.

M 1098, 1098, lambed 1887 ; *br.* T. Brown & Son, *s.* Favorite 109,
   d. M 82 by No. 9, 226,
   2nd d. Old Grey by No. 4, 219.

M 1099, 1099, lambed 1887 ; *br.* T. Brown & Son, *s.* Slasher 339,
   d. by Royal Kilburn 302.

M 1106, 1106, lambed 1887 ; *br.* T. Brown & Son, *s.* Favorite 109,

d. M 275 by Duke of York 94,

2nd d. by No. 1, 205.

M 1108, 1108, lambed 1887 ; *br.* T. Brown & Son, *s.* Paragon 268,

d. by Reserve 294.

M 1109, 1109, lambed 1887 ; *br.* T. Brown & Son, *s* Bayard 28,

d. by Reserve 294.

M 1112, 1112, lambed 1887 ; *br.* T. Brown & Son, *s.* Bayard 28,

d. M 366 by Gay Lad 126,

2nd d. by Son of Grey Nobleman 142.

M 1115, 1115, lambed 1887 ; *br.* T. Brown & Son, *s.* Paragon 268,

d. M 130 by Reserve 294,

2nd d. by Broadfield 54.

M 1123, 1123, lambed 1887 ; *br.* T. Brown & Son, *s.* Shylock 315,

d. by Royal Kilburn 302.

M 1125, 1125, lambed 1887 ; *br.* T. Brown & Son, *s.* Slasher 339,

d. by Reserve 294.

M 1126, 1126, lambed 1887 ; *br.* T Brown & Son, *s.* Slasher 339,

d. by Reserve 294.

M 1127, 1127, lambed 1887 ; *br.* T. Brown & Son, *s.* Shylock 315,

d. by Admiral 6.

M 1128, 1128, lambed 1887 ; *br.* T. Brown & Son, *s.* Shylock 315,

d. by Admiral 6.

M 1130, 1130, lambed 1887 ; *br.* T. Brown & Son, *s.* Spot 348,

d. M 252 by Example 102.

2nd d. by No. 8, 225.

M 1136, 1136, lambed 1887 ; *br.* T. Brown & Son, *s.* Eclipse 96,
       d. M 18 by Spot 348,
       2nd d. by Grey Lane 141.

M 1138, 1138, lambed 1887 ; *br.* T. Brown & Son, *s.* Slasher 339,
       d. M 334 by Beacon 34,
       2nd d. by No. 1, 205.

M 1140, 1140, lambed 1887 ; *br.* T. Brown & Son, *s.* Bayard 28,
       d. M 247 by Spot 348.
       2nd d. by Experience 106.

M 1142, 1142, lambed 1887 ; *br.* T. Brown & Son, *s.* Slasher 339,
       d. M 342 by Beacon 34,
       2nd d. by Royal Kilburn 302.

M 1143, 1143, lambed 1887 ; *br.* T. Brown & Son, *s.* Slasher 339,
       d. M 351 by Royal Kilburn 302,
       2nd d. by Reserve 294.

M 1144, 1144, lambed 1887 ; *br.* T. Brown & Son, *s.* Slasher 339,
       d. M 351 by Royal Kilburn 302,
       2nd d. by Reserve 294.

M 1147, 1147, lambed 1887 ; *br.* T. Brown & Son, *s.* Eclipse 96,

M 1148, 1148, lambed 1887 ; *br.* T. Brown & Son, *s.* Spot 348,
       d. by Admiral 6.

M 1149, 1149, lambed 1887 ; *br.* T. Brown & Son, *s.* Bayard 28,
       d. M 238, by Example 102,
       2nd d. by Lane's Old Sheep 166.

M 1152, 1152, lambed 1887 ; *br.* T. Brown & Son, *s.* Spot 348,
       d. M 114 by Spot 348,
       2nd d. by Broadfield 54.

M 1157, 1157, lambed 1887 ; *br.* T. Brown & Son, *s.* Citadel 70,
d. M 87 by Royal Derby 300,
2nd d. by Broadfield 54.

M 1159, 1159, lambed 1887 ; *br.* T. Brown & Son, *s.* Eclipse 96,
d. M 148 by Commodore 76,
2nd d. by Broadfield 54.

M 1164, 1164, lambed 1887 ; *br.* T. Brown & Son, *s.* Eclipse 96
d. M 599 by Lion 175,
2nd d. by Dick Allen 87.

M 1165, 1165, lambed 1887 ; *br.* T. Brown & Son, *s.* Shylock 315,
d. M 254 by Beacon 34,
2nd d. by Broadfield 54.

M 1166, 1166, lambed 1887 ; *br.* T. Brown & Son, *s.* Spot 348,
d. by Broadback 53.

M 1168, 1168, lambed 1887 ; *br.* T. Brown & Son, *s.* Paragon 268,
d. M 507 by Lion 175.
2nd d. by Admiral 6.

M 1169, 1169, lambed 1887 ; *br.* T. Brown & Son, *s.* Paragon 268,
d. M 268 by Example 102,
2nd d. by Blue Lane 48.

M 1171, 1171, lambed 1887 ; *br.* T. Brown & Son, *s.* Paragon 268,
d. M 406 by No. 20, 246,
2nd d. by Royal Kilburn 302.

M 1173, 1173, lambed 1887 ; *br.* T. Brown & Son, *s.* Shylock 315,
d. by Broadback 53.

M 1179, 1179, lambed 1887 ; *br.* T. Brown & Son, *s.* Shylock 315,
d. by Broadback 53.

M 1180, 1180, lambed 1887 ; *br.* T. Brown & Son, *s.* Eclipse 96,
　　　　d. M 633 by No. 2, 211,
　　　　2nd d. M 156 by Spot 348,
　　　　3rd d. by Marham Blackfoot 185.

M 1182, 1182, lambed 1887 ; *br.* T. Brown & Son, *s.* Spot 348,
　　　　d. M 472 by Kent 163,
　　　　2nd d. by Broadfield 54.

M 1185, 1185, lambed 1887 ; *br.* T. Brown & Son, *s.* Bayard 28,
　　　　d. M 625 by Spot 348,
　　　　2nd d. by General 129.

M 1193, 1193, lambed 1887 ; *br.* T. Brown & Son, *s.* Cavalier 66,
　　　　d. by Broadback 53.

M 1198, 1198, lambed 1887 ; *br.* T. Brown & Son, *s.* Paragon 268,
　　　　d. M 506 by Kent 163,
　　　　2nd d. by Sultan 352.

M 1205, 1205, lambed 1887 ; *br.* T. Brown & Son, *s.* Shylock 315,
　　　　d. by Garne's Blackfoot 123.

M 1207, 1207, lambed 1887 ; *br.* T. Brown & Son, *s.* Eclipse 96,
　　　　d. M 628 by No. 2, 211,
　　　　2nd d. M 92 by Royal Derby 300,
　　　　3rd d. by Marham Blackfoot 185.

M 1212, 1212, lambed 1887 ; *br.* T. Brown & Son, *s.* Shylock 315,
　　　　d. M 304 by Spot 348,
　　　　2nd d. by Broadback 53.

M 1216, 1216, lambed 1887 ; *br.* T. Brown & Son, *s.* Paragon 268,
　　　　d. M 585 by Example 102,
　　　　2nd d. by Royal Kilburn 302.

M 1219, 1219, lambed 1887 ; *br.* T. Brown & Son, *s.* Paragon 268,
d. M 608 by Singleton 324,
2nd d. by Royal Kilburn 302.

M 1223, 1223, lambed 1887 ; *br.* T. Brown & Son, *s.* Slasher 339,
d. M 636 by Beacon 34,
2nd d. M 121 by Spot 348,
3rd d. by Reserve 294.

M 1224, 1224, lambed 1887 ; *br.* T. Brown & Son, *s.* Spot 348,
d. M 504 by Kent 163,
2nd d. by Marham Blackfoot 185.

M 1226, 1226, lambed 1887 ; *br.* T. Brown & Son, *s.* Slasher 339,
d. M 516 by Spot 348,
2nd d. by Broadfield 54.

M 1230, 1230, lambed 1887 ; *br.* T. Brown & Son, *s.* Slasher 339,
d. M 430 by No. 20, 246,
2nd d. by Grey Lane 141.

M 1231, 1231, lambed 1887 ; *br.* T. Brown & Son, *s.* Eclipse 96,
d. by Reserve 294.

M 1235, 1235, lambed 1887 ; *br.* T. Brown & Son, *s.* Eclipse 96,
d. M 284 by Spot 348,
2nd d. by Broadfield 54.

M 1239, 1239, lambed 1887 ; *br.* T. Brown & Son, *s.* Eclipse 96,
d. by Reserve 294.

M 1241, 1241, lambed 1887 ; *br.* T. Brown & Son, *s.* Slasher 339,
d. by Experience 106.

M 1250, 1250, lambed 1887 ; *br.* T. Brown & Son, *s.* Shylock 315,
d. by Experience 106.

M 1251, 1251, lambed 1887 ; *br.* T. Brown & Son, *s.* Shylock 315,
    d. by Experience 106.

M 1253, 1253, lambed 1887 ; *br.* T. Brown & Son, *s.* Slasher 339,
    d. M 572 by Surprise 353,
    2nd d. by Royal Kilburn 302.

M 1255, 1255, lambed 1887 ; *br.* T. Brown & Son, *s.* Shylock 315,
    d. M 614 by Beacon 34,
    2nd d. by Royal Kilburn 302.

M 1257, 1257, lambed 1887 ; *br.* T. Brown & Son, *s.* Slasher 339,
    d. by Broadback 53.

M 1261, 1261, lambed 1887 ; *br.* T. Brown & Son, *s.* Eclipse 96,
    d. M 348 by Example 102,
    2nd d. by Broadfield 54.

M 1265, 1265, lambed 1887 ; *br.* T. Brown & Son, *s.* Citadel 70,
    d. M 413 by Commodore 76,
    2nd d. by Grey Son of Broadfield 143.

M 1266, 1266, lambed 1887 ; *br.* T. Brown & Son, *s.* Eclipse 96,
    d. by Admiral 6.

M 1268, 1268, lambed 1887 ; *br.* T. Brown & Son, *s.* Slasher 339,
    d. M 704 by Surprise 353,
    2nd d. by Experience 106.

M 1269, 1269, lambed 1887 ; *br.* T. Brown & Son, *s.* Slasher 339,
    d. M 496 by Example 102,
    2nd d. by Reserve 294.

M 1273, 1273, lambed 1887 ; *br.* T. Brown & Son, *s.* Eclipse 96,
    d. M 523 by Kent 163,
    2nd d. by No. 1, 204.

M 1275, 1275, lambed 1887 ; *br.* T. Brown & Son, *s.* Shylock 315,
         d. M 184 by Commodore 76,
         2nd d. by Little Doctor 177.

M 1278, 1278, lambed 1887 ; *br.* T. Brown & Son,
         s. Son of Gay Lad 344,
         d. by Admiral 6.

M 1280, 1280, lambed 1887 ; *br.* T. Brown & Son,
         s. Son of Gay Lad 344,
         d. The Pretty Ewe.

M 1281, 1281, lambed 1887 ; *br.* T. Brown & Son,
         s. Son of Gay Lad 344,
         d. The Pretty Ewe.

M 1283, 1283, lambed 1888 ; *br.* T. Brown & Son, *s.* Granite 136,
         d. M 13 by Reserve 294,
         2nd d. by Royal Kilburn 302.

M 1284, 1284, lambed 1888 ; *br.* T. Brown & Son, *s.* Abbot 3,
         d. M 280 by Gay Lad 126,
         2nd d. by Dick Allen 87.

M 1286, 1286, lambed 1888 ; *br.* T. Brown & Son, *s.* Primate 278,
         d. by Example 102.

M 1287, 1287, lambed 1888 ; *br.* T. Brown & Son, *s.* Granite 136,
         d. M 88 by Commodore 76,
         2nd d. by Marham Blackfoot 185.

M 1288, 1288, lambed 1888 ; *br.* T. Brown & Son, *s.* Granite 136,
         d. M 172 by Commodore 76,
         2nd d. by Grey Lane 141.

M 1290, 1290, lambed 1888 ; *br.* T. Brown & Son, *s.* Lancer 164,
d. M 803 by Favorite 109,
2nd d. by Reserve 294.

M 1298, 1298, lambed 1888 ; *br.* T. Brown & Son, *s.* Granite 136,
d. by Royal Kilburn 302.

M 1301, 1301, lambed 1888 ; *br.* T. Brown & Son, *s.* Bernard 37,
d. M 496 by Example 102,
2nd d. by Reserve 294.

M 1303, 1303, lambed 1888 ; *br.* T. Brown & Son, *s.* Bernard 37,
d. M 771 by Excelsior 103,
2nd d. M 5 by No. 20, 246,
3rd d. by Ensign 99.

M 1304, 1304, lambed 1888 ; *br.* T. Brown & Son, *s.* Abbot 3,
d. M 625 by Spot 348,
2nd d. by General 129.

M 1305, 1305, lambed 1888 ; *br.* T. Brown & Son, *s.* Comet 75,
d. M 1006 by Garnet 124,
2nd d. M 315 by Kent 163,
3rd d. by No. 42, 256.

M 1318, 1318, lambed 1888 ; *br.* T. Brown & Son, *s.* Granite 136,
d. by Marham Blackfoot 185.

M 1319, 1319, lambed 1888 ; *br.* T. Brown & Son, *s.* Granite 136,
d. M 532 by Lion 175,
2nd d. by Marham Blackfoot 185.

M 1322, 1322, lambed 1888 ; *br.* T. Brown & Son, *s.* Comet 75,
d. M 780 by Sturdy 350,
2nd d. by Admiral 6.

M 1323, 1323, lambed 1888 ; *br.* T. Brown & Son, *s.* Primate 278.

M 1324, 1324, lambed 1888 ; *br.* T. Brown & Son, *s.* Paragon 268,
d. M 788 by Favorite 109,
2nd d. by Dick Allen 87.

M 1325, 1325, lambed 1888 ; *br.* T. Brown & Son, *s.* Granite 136,
d. M 365 by Duke of York 94.

M 1327, 1327, lambed 1888 ; *br.* T. Brown & Son, *s.* Primate 278,
d. M 399 by Haylock 146,
2nd d. by Admiral 6.

M 1328, 1328, lambed 1888 ; *br.* T. Brown & Son, *s.* Granite, 136,
d. M 142 by Quality 289.
2nd d. by Grey Nobleman 142.

M 1331, 1331, lambed 1888 ; *br.* T. Brown & Son, *s.* Abbot 3,
d. M 340 by Haylock 146,
2nd d. by Lane's Old Sheep 166.

M 1332, 1332, lambed 1888 ; *br.* T. Brown & Son, *s.* Granite 136,
d. by Reserve, 294.

M 1341, 1341, lambed 1888 ; *br.* T. Brown & Son, *s.* Granite 136,
d. M 241 by Haylock 146,
2nd d. by Royal Kilburn 302.

M 1342, 1342, lambed 1888 ; *br.* T. Brown & Son, *s.* Primate 278,
d. by Admiral 6.

M 1349, 1349, lambed 1888 ; *br.* T. Brown & Son, *s.* Granite 136,
d. M 364 by Duke of York 94.

M 1351, 1351, lambed 1888 ; *br.* T. Brown & Son, *s.* Lancer 164,
          d. M 827 by Favorite 109,
          2nd d. M 156 by Spot 348,
          3rd d. by Marham Blackfoot 185.

M 1352, 1352, lambed 1888 ; *br.* T. Brown & Son, *s.* Granite 136,
          d. M 206 by Royal Kilburn 302,
          2nd d. by Broadfield 54.

M 1354, 1354, lambed 1888 ; *br.* T. Brown & Son, *s.* Abbot 3.

M 1355, 1355, lambed 1888 ; *br.* T. Brown & Son, *s.* Lancer 164,
          d. M 860 by Excelsior 103,
          2nd d. by Admiral 6.

M 1356, 1356, lambed 1888 ; *br.* T. Brown & Son, *s.* Lancer 164,
          d. M 865 by Spot 348,
          2nd d. by Reserve 294.

M 1359, 1359, lambed 1888 ; *br.* T. Brown & Son, *s.* Bernard 37,
          d. M 793 by Favorite 109,
          2nd d. by Example 102.

M 1360, 1360, lambed 1888 ; *br.* T. Brown & Son, *s.* Abbot 3,
          d. M 519 by Example 102,
          2nd d. by No. 2, 210.

M 1363, 1363, lambed 1888 ; *br.* T. Brown & Son, *s.* Paragon 268,
          d. M 705 by No. 2, 211,
          2nd d. M 106 by Quality 289,
          3rd d. by Reserve 294.

M 1364, 1364, lambed 1888 ; *br.* T. Brown & Son, *s.* Paragon 268,
          d. M 555 by Singleton 324,
          2nd d. by Broadfield 54.

M 1365, 1365, lambed 1888 ; *br.* T. Brown & Son, *s.* Granite 136,
      d. M 128 by Gay Lad 126,
      2nd d. by Grey Lane 141.

M 1366, 1366, lambed 1888 ; *br.* T. Brown & Son, *s.* Granite 136,
      d. by Reserve 294.

M 1368, 1368, lambed 1888 ; *br.* T. Brown & Son, *s.* Lancer 164,
      d. M 864 by Spot 348,
      2nd d. by Reserve 294.

M 1380, 1380, lambed 1888 ; *br.* T. Brown & Son, *s.* Granite 136,
      d. M 304 by Spot 348,
      2nd d. by Broadback 53.

M 1384, 1384, lambed 1888 ; *br.* T. Brown & Son, *s.* Granite 136,
      d. M 130 by Reserve 294.
      2nd d. by Broadfield 54.

M 1386, 1386, lambed 1888 ; *br.* T. Brown & Son, *s.* Lancer 164,
      d. M 657 by Commodore 76,
      2nd d. by Reserve 294.

M 1388, 1388, lambed 1888 ; *br.* T. Brown & Son, *s.* Comet 75,
      d. Pedley's 4.

M 1389, 1389, lambed 1888 ; *br.* T. Brown & Son, *s.* Granite 136,
      d. M 429 by No. 20, 246,
      2nd d. by Royal Kilburn 302.

M 1397, 1397, lambed 1888 ; *br.* T. Brown & Son, *s.* Lancer 164,
      d. M 773 by Ronald 297,
      2nd d. M 305 by Spot 348,
      3rd d. by Reserve 294.

M 1400, 1400, lambed 1888; *br.* T. Brown & Son, *s.* Comet 75,

  d. M 862 by Excelsior 103,

  2nd d. by Admiral 6.

M 1401, 1401, lambed 1888; *br.* T. Brown & Son, *s.* Abbot 3,

  d. M. 598 by Example 102,

  2nd d. by Garne's Blackfoot 123.

M 1402, 1402, lambed 1888; *br.* T. Brown & Son, *s.* Abbot 3,

  d. M 598 by Example 102,

  2nd d. by Garne's Blackfoot 123.

M 1404, 1404, lambed 1888; *br.* T. Brown & Son, *s.* Granite 136,

  d. M 29 by Gay Lad 126,

  2nd d. by General 129.

M 1405, 1405, lambed 1888; *br.* T. Brown & Son, *s.* Granite 136,

  d. M 145 by Commodore 76,

  2nd d. by No. 42, 256.

M 1406, 1406, lambed 1888; *br.* T. Brown & Son, *s.* Granite 136,

  d. M 145 by Commodore 76,

  2nd d. by No. 42, 256.

M 1407, 1407, lambed 1888; *br.* T. Brown & Son, *s.* Granite 136,

  d. M 106 by Quality 289,

  2nd d. by Reserve 294.

M 1410, 1410, lambed 1888; *br.* T. Brown & Son, *s.* Granite 136,

  d. by Admiral 6.

M 1416, 1416, lambed 1888; *br.* T. Brown & Son, *s.* Percy 270,

  d. M 587 by Surprise 353,

  2nd d. by Grey Nobleman 142.

M 1418, 1419, lambed 1888 ; *br.* T. Brown & Son, *s.* Abbot 3,
d. M 327 by Royal Kilburn 302,
2nd d. by Broadfield 54.

M 1420, 1420, lambed 1888 ; *br.* T. Brown & Son, *s.* Granite 136,
d. M 150 by Gay Lad 126,
2nd d. by Reserve 294.

M 1422, 1422, lambed 1888 ; *br.* T. Brown & Son, *s.* Bernard 37,
d. M 199 by Commodore 76,
2nd d. by No. 42, 256.

M 1423, 1423, lambed 1888 ; *br.* T. Brown & Son, *s.* Primate 278,
d. M 313 by Commodore 76,
2nd d. by Broadfield 54.

M 1428, 1428, lambed 1888 ; *br.* T. Brown & Son, *s.* Conrad 77,
d. M 339 by Commodore 76,
2nd d. by Broadfield 54.

M 1429, 1429, lambed 1888 ; *br.* T. Brown & Son, *s.* Abbot 3,
d. M 166 by Commodore 76,
2nd d. by No. 8, 225.

M 1430, 1430, lambed 1888 ; *br.* T. Brown & Son, *s.* Abbot 3,
d. M 166 by Commodore 76,
2nd d. by No. 8, 225.

M 1432, 1432, lambed 1888 ; *br.* T. Brown & Son, *s.* Granite 136,
d. by Royal Kilburn 302.

M 1436, 1436, lambed 1888 ; *br.* T. Brown & Son, *s.* Abbot 3,
d. by Sultan 352.

M 1437, 1437, lambed 1888 ; *br.* T. Brown & Son, *s.* Abbot 3,
d. M 305 by Spot 348,
2nd d. by Reserve 294.

M 1438, 1438, lambed 1888 ; *br.* T. Brown & Son, *s.* Bernard 37,

    d. M 504 by Kent 163,

    2nd d. by Marham Blackfoot 185.

M 1442, 1442, lambed 1888 ; *br.* T. Brown & Son, *s.* Granite 136,

    d. M 473 by Lion 175,

    2nd d. by Reserve 294.

M 1446, 1446, lambed 1888 ; *br.* T. Brown & Son, *s.* Granite 136,

    d. M 545 by Surprise 353,

    2nd d. by Grey Nobleman 142.

M 1447, 1447, lambed 1888 ; *br.* T. Brown & Son, *s.* Granite 136,

    d. Crop Ear by K. B. 159.

M 1448, 1448, lambed 1888 ; *br.* T. Brown & Son, *s.* Granite 136,

    d. M 4 by Gay Lad 126,

    2nd d. by Boreas 51.

M 1449, 1449, lambed 1888 ; *br.* T. Brown & Son, *s.* Bernard 37,

    d. M 310 by Experience 106,

    2nd d. by Garne's Blackfoot 123.

M 1450, 1450, lambed 1888 ; *br.* T. Brown & Son, *s.* Bernard 37,

    d. M 310 by Experience 106,

    2nd d. by Garne's Blackfoot 123.

M 1451, 1451, lambed 1888 ; *br.* T. Brown & Son, *s.* Lancer 164,

    d. M 247 by Spot 348,

    2nd d. by Experience 106.

M 1452, 1452, lambed 1888 ; *br.* T. Brown & Son, *s.* Granite 136,

    d. M 131 by Spot 348,

    2nd d. by Reserve 294.

M 1454, 1454, lambed 1888 ; *br.* T. Brown & Son, *s.* Granite 136,
d. by Broadfield 54.

M 1457, 1457, lambed 1888 ; *br.* T. Brown & Son, *s.* Bernard 37,
d. M 840 by Spot 348,
2nd d. by Reserve 294.

M 1459, 1459, lambed 1888 ; *br.* T. Brown & Son, *s.* Lancer 164,
d. M 663 by Surprise 353.

M 1461, 1461, lambed 1888 ; *br.* T. Brown & Son, *s.* Abbot 3,
d. M 859 by Favorite 109,
2nd d. by Broadfield 54.

M 1475, 1475, lambed 1888 ; *br.* T. Brown & Son, *s.* Primate 278,
d. M 114 by Spot 348,
2nd d. by Broadfield 54.

M 1477, 1477, lambed 1888 ; *br.* T. Brown & Son, *s.* Claudian 71,
d. M 174 by Experience 106,
2nd d. by Grey Nobleman 142.

M 1479, 1479, lambed 1888 ; *br.* T. Brown & Son, *s.* Lancer 164,
d. M 660 by Beacon 34.

M 1480, 1480, lambed 1888 ; *br.* T. Brown & Son, *s.* Claudian 71,
d. M 783 by Spot 348,
2nd d. by Experience 106.

M 1481, 1481, lambed 1888 ; *br.* T. Brown & Son, *s.* Granite 136,
d. M 430 by No. 20, 246,
2nd d. by Grey Lane 141.

M 1482, 1482, lambed 1888 ; *br.* T. Brown & Son, *s.* Granite 136,
d. M 430 by No. 20, 246,
2nd d. by Grey Lane 141.

M 1483, 1483, lambed 1888 ; *br.* T. Brown & Son, *s.* Percy 270,
    d. by Reserve 294.

M 1485, 1485, lambed 1888 ; *br.* T. Brown & Son, *s.* Percy 270,
    d. M 485 by Lion 175,
    2nd d. by No. 42, 256.

M 1491, 1491, lambed 1888 ; *br.* T. Brown & Son, *s.* Lancer 164,
    d. M 925 by Sturdy 350,
    2nd d. by Experience 106.

M 1492, 1492, lambed 1888 ; *br.* T. Brown & Son, *s.* Bernard 37,
    d. M 520 by Example 102.
    2nd d. by No. 2, 210.

M 1493, 1493, lambed 1888 ; *br.* T. Brown & Son, *s.* Lancer 164,
    d. M 322 by Haylock 146,
    2nd d. by Sultan 352.

M 1496, 1496, lambed 1888 ; *br.* T. Brown & Son, *s.* Claudian 71,
    d. M 920 by Paragon 268,
    2nd d. M 216 by Son of No. 8, 347,
    3rd d. by General 129.

M 1500, 1500, lambed 1888 ; *br.* T. Brown & Son, *s.* Oliver 264,
    d. M 684 by Beacon 54,
    2nd d. by Admiral 6.

M 1501, 1501, lambed 1888 ; *br.* T. Brown & Son, *s.* Abbot 3,
    d. M 600 by Singleton 324,
    2nd d. by No. 8, 123.

M 1509, 1509, lambed 1888 ; *br.* T. Brown & Son, *s.* Claudian 71,
    d. Pedley's 12.

M 1511, 1511, lambed 1888 ; *br.* T. Brown & Son, *s.* Colin 72,

      d. M 964 by Garnet 124,

      2nd d. M 134 by Spot 348,

      3rd d. by No. 2, 210.

M 1532, 1532, lambed 1888 ; *br.* T. Brown & Son, *s.* Cadet 62,

      d. by Experience 106.

M 1536, 1536, lambed 1888 ; *br.* T. Brown & Son, *s.* Bernard 37,

      d. M 1 by Royal Derby 300,

      2nd d. by Lane's Old Sheep 166.

M 2a,   2, lambed 1889 ; *br.* T. Brown & Son, *s.* Leonard 173,

      d. M 1125 by Slasher 339,

      2nd d. by Reserve 294.

M 5a,   5, lambed 1889 ; *br.* T. Brown & Son, *s.* Fyfield 122,

      d. M 821 by Favorite 109,

      2nd d. by Royal Kilburn 302.

M 6a,   6, lambed 1889 ; *br.* T. Brown & Son, *s.* Leonard 173,

      d. M 771 by Excelsior 103,

      2nd d. M 5 by No. 20, 246,

      3rd d. by Ensign 99.

M 9a,   9, lambed 1889 ; *br.* T. Brown & Son, *s.* Granite 136,

      d. M 649 by Surprise 353,

      2nd d. by Broadfield 54.

M 10a, 10, lambed 1889 ; *br.* T. Brown & Son, *s.* Ajax 8,

      d. M 478 by Spot 348,

      2nd d. by Reserve 294.

M 17*a*, 17, lambed 1889 ; *br.* T. Brown & Son, *s.* Ajax 8,
d. M 1171 by Paragon 268,
2*nd d.* M 406 by No. 20, 246,
3*rd d.* by Royal Kilburn 302.

M 20*a*, 20, lambed 1889 ; *br.* T. Brown & Son, *s.* Leonard 173,
d. by Rear-Admiral 292.

M 22*a*, 22, lambed 1889 ; *br.* T. Brown & Son, *s.* Oscar 265,
d. M 759 by Sturdy 350,
2*nd d.* by No. 8, 225.

M 25*a*, 25, lambed 1889 ; *br.* T. Brown & Son, *s.* Ajax 8,
d. M 182 by Reserve 294,
2*nd d.* by Royal Kilburn 302.

M 29*a*, 29, lambed 1889 ; *br.* T. Brown & Son, *s.* Colin 72,
d. M 423 by No. 20, 246,
2*nd d.* by General 129.

M 30*a*, 30, lambed 1889 ; *br.* T. Brown & Son, *s.* Fyfield 122,
d. M 1268 by Slasher 336,
2*nd d.* M 704 by Surprise 353,
3*rd d.* by Experience 106.

M 31*a*, 31, lambed 1889 ; *br.* T. Brown & Son, *s.* Forester 113,
d. Pedley's 2.

M 34*a*, 34, lambed 1889 ; *br.* T. Brown & Son, *s.* Leonard 173,
d. M 145 by Commodore 76,
2*nd d.* by No. 42, 256.

M 35*a*, 35, lambed 1889 ; *br.* T. Brown & Son, *s.* Leonard 173,
d. M 145 by Commodore 76,
2*nd d.* by No. 42, 256.

M 39a, 39, lambed 1889 ; *br.* T. Brown & Son, *s.* Fyfield 122,
     d. M 428 by No. 20, 246,
     2*nd d.* by Broadfield 54.

M 43a, 43, lambed 1889 ; *br.* T. Brown & Son, *s* Granite 136,
     d. M 142 by Quality 289,
     2*nd d.* by Grey Nobleman 142.

M 44a, 44, lambed 1889 ; *br.* T. Brown & Son, *s.* Granite 136,
     d. M 142 by Quality 289,
     2*nd d.* by Grey Nobleman 142.

M 45a, 45, lambed 1889 ; *br.* T. Brown & Son, *s.* Granite 136,
     d. M 142 by Quality 289,
     2*nd d.* by Grey Nobleman 142.

M 49a, 49, lambed 1889 ; *br.* T. Brown & Son, *s.* Oscar 265,
     d. M 397 by Duke of York 94,
     2*nd d.* by K. B. 159.

M 51a, 51, lambed 1889 ; *br.* T. Brown & Son, *s.* Fyfield 122,
     d. M 465 by Lion 175,
     2*nd d.* by No. 42, 256.

M 55a, 55, lambed 1889 ; *br.* T. Brown & Son, *s.* Ajax 8,
     d. M 6 by Gay Lad 126,
     2*nd d.* by No. 42, 256.

M 58a, 58, lambed 1889 ; *br.* T. Brown & Son, *s.* Granite 136,
     d. by Reserve 294.

M 60a, 60, lambed 1889 ; *br.* T. Brown & Son, *s.* Leonard 173,
     d. M 275 by Duke of York 94,
     2*nd d.* by No. 1, 205.

M 63*a*, 63, lambed 1889 ; *br.* T. Brown & Son, *s.* Granite 136,

> *d.* M 208 by Reserve 294,
>
> 2*nd d.* by Grey Nobleman 142.

M 64*a*, 64, lambed 1889 ; *br.* T. Brown & Son, *s.* Granite 136,

> *d.* M 364 by Duke of York 94.

M 65*a*, 65, lambed 1889 ; *br.* T. Brown & Son, *s.* Forester 113,

> *d.* M 794 by Ronald 297,
>
> 2*nd d.* by Sultan 352.

M 68*a*, 68, lambed 1889 ; *br.* T. Brown & Son, *s.* Colin 72,

> *d.* M 32 by Gay Lad 126,
>
> 2*nd d.* by Boreas 51.

M 69*a*, 69, lambed 1889 ; *br.* T. Brown & Son, *s.* Ajax 8,

> *d.* M 365 by Duke of York 94.

M 71*a*, 71, lambed 1889 ; *br.* T. Brown & Son, *s.* Oscar 265,

> *d.* M 281 by Beacon 34,
>
> 2*nd d.* by No. 8, 225.

M 72*a*, 72, lambed 1889 ; *br.* T. Brown & Son, *s.* Ajax 8,

> *d.* M 685 by Beacon 34,
>
> 2*nd d.* by Admiral 6.

M 75*a*, 75, lambed 1889 ; *br.* T. Brown & Son, *s.* Eric 101,

> *d.* M 498 by Kent 163,
>
> 2*nd d.* by Marham Blackfoot 185.

M 76*a*, 76, lambed 1889 ; *br.* T. Brown & Son, *s.* Ajax 8,

> *d.* M 245 by Spot 348,
>
> 2*nd d.* by Royal Kilburn 302.

M 77*a*, 77, lambed 1889 ; *br.* T. Brown & Son, *s.* Ajax 8,
      d. M 245 by Spot 348,
      2*nd d.* by Royal Kilburn 302.

M 78*a*, 78, lambed 1889; *br.* T. Brown & Son, *s.* Ajax 8,
      d. M 1159 by Eclipse 96,
      2*nd d.* M 148 by Commodore 76.
      3*rd d.* by Broadfield 54.

M 80*a*, 80, lambed 1889; *br.* T. Brown & Son, *s.* Oscar 265,
      d. M 305 by Spot 348,
      2*nd d.* by Reserve 294.

M 81*a*, 81, lambed 1889 ; *br.* T. Brown & Son, *s.* Ajax 8,
      d. M 304 by Spot 348,
      2*nd d.* by Broadback 53.

M 82*a*, 82, lambed 1889 ; *br.* T. Brown & Son, *s.* Ajax 8,
      d. by Reserve 294.

M 83*a*, 83, lambed 1889 ; *br.* T. Brown & Son, *s.* Ajax 8,
      d. M 252 by Example 102,
      2*nd d.* by No. 8, 225.

M 85*a*, 85, lambed 1889 ; *br.* T. Brown & Son, *s.* Ajax 8,
      d. M 339 by Commodore 76,
      2*nd d.* by Broadfield 54.

M 86*a*, 86, lambed 1889 ; *br.* T. Brown & Son, *s.* Oscar 265,
      d. M 481 by Kent 163,
      2*nd d.* by Marham Blackfoot 185.

M 87*a*, 87, lambed 1889 ; *br.* T. Brown & Son, *s.* Oscar 265,
      d. M 481 by Kent 163,
      2*nd d.* by Marham Blackfoot 185.

M 91*a*, 91, lambed 1889 ; *br.* T. Brown & Son, *s.* Ajax 8,

         d. M 247 by Spot 348,

         2nd d. by Experience 106.

M 95*a*, 95, lambed 1889 ; *br.* T. Brown & Son, *s.* Eric 101,

         d. M 623 by No. 2, 211,

         2nd d. M 94 by Commodore 76,

         3rd d. by Broadfield 54.

M 98*a*, 98, lambed 1889 ; *br.* T. Brown & Son, *s.* Ajax 8,

         d. M 845 by Sturdy 350,

         2nd d. M 27 by No 20, 246,

         3rd d. by No. 3, 2nd, 218,

M 100*a*, 100, lambed 1889 ; *br.* T. Brown & Son, *s.* Granite 136,

         d. M 92 by Royal Derby 300,

         2nd d. by Marham Blackfoot 185.

M 102*a*, 102, lambed 1889 ; *br.* T. Brown & Son, *s.* Fyfield 122,

         d. M 760 by Garnet 124,

         2nd d. by Royal Kilburn 302.

M 104*a*, 104, lambed 1889 ; *br.* T. Brown & Son, *s.* Ajax 8,

         d. M 830 by Ronald 297,

         2nd d. by Broadback 53.

M 105*a*, 105, lambed 1889 ; *br.* T. Brown & Son, *s.* Ajax 8,

         d. M 94 by Commodore 76,

         2nd d. by Broadfield 54.

M 106*a*, 106, lambed 1889 ; *br.* T. Brown & Son, *s.* Ajax 8,

         d. M 836 by Paragon 268,

         2nd d. M 173 by Son of K. B. 346,

         3rd d. by Morton 190.

M 107*a*, 107, lambed 1889 ; *br.* T. Brown & Son, *s.* Ajax 8,
           d. M 976 by Paragon 268.
           2nd d. M 172 by Commodore 76,
           3rd d. by Grey Lane 141.

M 109*a*, 109, lambed 1889 ; *br.* T. Brown & Son, *s.* Granite 136,
           d. M 200 by Quality 289,
           2nd d. by Grey Nobleman 142.

M 110*a*, 110, lambed 1889 ; *br.* T. Brown & Son, *s.* Leonard 173,
           d. M 112 by Royal Derby 300,
           2nd d. by Little Doctor 177.

M 112*a*, 112, lambed 1889 ; *br.* T. Brown & Son, *s.* Ajax 8,
           d. M 613 by Example 102,
           2nd d. by Reserve 294.

M 114*a*, 114, lambed 1889 ; *br.* T. Brown & Son, *s.* Fyfield 122,
           d. M 1006 by Garnet 124,
           2nd d. M 315 by Kent 163,
           3rd d. by No. 42, 256.

M 119*a*, 119, lambed 1889 ; *br.* T. Brown & Son, *s.* Granite 136,
           d. by Broadback 53.

M 122*a*, 122, lambed 1889 ; *br.* T. Brown & Son, *s.* Fyfield 122,
           d. M 861 by Favorite 109,
           2nd d. M 41 by Reserve 294,
           3rd d. by Blue Lane 48.

M 123*a*, 123, lambed 1889 ; *br.* T. Brown & Son, *s.* Fyfield 122,
           d. M 940 by Favorite 109,
           2nd d. M 89 by No. 9, 226,
           3rd d. by Royal Kilburn 302.

M 124*a*, 124, lambed 1889 ; *br.* T. Brown & Son, *s.* Leonard 173,
      d. M 406 by No. 20, 246,
      2*nd d.* by Royal Kilburn 302.

M 129*a*, 129, lambed 1889 ; *br.* T. Brown & Son, *s.* Ajáx 8,
      d. M 815 by Paragon 268.
      2*nd. d.* by Broadfield 54.

M 130*a*, 130, lambed 1889 ; *br.* T. Brown & Son, *s.* Leonard 173,
      d. M 156 by Spot 348,
  r   2*nd d.* by Marham Blackfoot 185.

M 131*a*, 131, lambed 1889 ; *br.* T. Brown & Son, *s.* Ajax 8,
      d. M 723 by Example 102,
      2*nd d.* M 182 by Reserve 294,
      3*rd d.* by Royal Kilburn 302.

M 132*a*, 132, lambed 1889 ; *br.* T. Brown & Son, *s.* Ajax 8,
      d. M 723 by Example 102,
      2*nd d.* M 182 by Reserve 294,
      3*rd d.* by Royal Kilburn 302.

M 133*a*, 133, lambed 1889 ; *br.* T. Brown & Son, *s.* Granite 136,
      d. M 4 by Gay Lad 126,
      2*nd d.* by Boreas 51.

M 134*a*, 134, lambed 1889 ; *br.* T. Brown & Son, *s.* Granite 136,
      d. M 4 by Gay Lad 126,
      2*nd d.* by Boreas 51.

M 139*a*, 139, lambed 1889 ; *br.* T. Brown & Son, *s.* Leonard 173,
      d. M 824 by Excelsior 103,
      2*nd d.* by Broadback 53.

M 140*a*, 140, lambed 1889 ; *br.* T. Brown & Son, *s.* Forester 113,
d. M 951 by Garnet 124,
2*nd d.* by Broadfield 54.

M 141*a*, 141, lambed 1889 ; *br.* T. Brown & Son, *s.* Eric 101,
d. M 817 by Sturdy 350,
2*nd d.* by No. 2, 210.

M 145*a*, 145, lambed 1889 ; *br.* T. Brown & Son, *s.* Leonard 173,
d. M 12 by Spot 348,
2*nd d.* by No. 4, 219.

M 150*a*, 150, lambed 1889 ; *br.* T. Brown & Son, *s.* Granite 136,
d. M 479 by Surprise 353,
2*nd d.* by Reserve 294.

M 151*a*, 151, lambed 1889 ; *br.* T. Brown & Son, *s.* Ajax 8,
d. M 301 by Experience 106,
2*nd d.* by Reserve 294.

M 152*a*, 152, lambed 1889 ; *br.* T. Brown & Son, *s.* Ajax 8,
d. M 301 by Experience 106,
2*nd d.* by Reserve 294.

M 157*a*, 157, lambed 1889 ; *br.* T. Brown & Son, *s.* Forester 113,
d. by Reserve 294.

M 160*a*, 160, lambed 1889 ; *br.* T. Brown & Son, *s.* Forester 113,
d. M 372 by Duke of York 94,
2*nd d.* by Admiral 6.

M 161*a*, 161, lambed 1889 ; *br.* T. Brown & Son, *s.* Oscar 265,
d. M 388 by Spot 348,
2*nd d.* by Broadback 53.

M 162*a*, 162, lambed 1889 ; *br.* T. Brown & Son, *s.* Ajax 8,

        d. M 532 by Lion 175,

        2nd d. by Marham Blackfoot 185.

M 167*a*, 167, lambed 1889 ; *br.* T. Brown & Son, *s.* Fyfield 122,

        d. M 1149 by Bayard 28,

        2nd d. M 238 by Example 102,

        3rd d. by Lane's Old Sheep 166.

M 169*a*, 169, lambed 1889 ; *br.* T. Brown & Son, *s.* Eric 101,

        d. M 233 by Beacon 34,

        2nd d. by No. 42, 256.

M 171*a*, 171, lambed 1889 ; *br.* T. Brown & Son, *s.* Granite 136,

        d. M 1180 by Eclipse 96,

        2nd d. M 633 by No. 2, 211,

        3rd d. M 156 by Spot 348,

        4th d. by Marhan Blackfoot 185.

M 173*a*, 173, lambed 1889 ; *br.* T. Brown & Son, *s.* Leonard 173,

        d. M 184 by Commodore 76,

        2nd d. by Little Doctor 177.

M 175*a*, 175, lambed 1889 ; *br.* T. Brown & Son, *s.* Fyfield 122,

        d. M 130 by Reserve 294,

        2nd d. by Broadfield 54.

M 178*a*, 178, lambed 1889 ; *br.* T. Brown & Son. *s.* Eric 101,

        d. M 254 by Beacon 34,

        2nd d. by Broadfield 54.

.M 179*a*, 179, lambed 1889 ; *br.* T. Brown & Son, *s.* Leonard 173,

        d. M 348 by Example 102,

        2nd d. by Broadfield 54.

M 180*a*, 180, lambed 1889 ; *br.* T. Brown & Son, *s.* Leonard 173,
d. M 348 by Example 102,
2nd d. by Broadfield 54.

M 182*a*, 182, lambed 1889 ; *br.* T. Brown & Son, *s.* Leonard 173,
d. M 148 by Commodore 76,
2nd d. by Broadfield 54.

M 187*a*, 187, lambed 1889 ; *br.* T. Brown & Son, *s.* Fyfield 122,
d. M 1231 by Eclipse 96,
2nd d. by Reserve 294.

M 194*a*, 194, lambed 1889 ; *br.* T. Brown & Son, *s.* Leonard 173,
d. M 87 by Royal Derby 300.
2nd d. by Broadfield 54.

M 195*a*, 195, lambed 1889 ; *br.* T. Brown & Son, *s.* Leonard 173,
d. M 87 by Royal Derby 300,
2nd d. by Broadfield 54.

M 196*a*, 196, lambed 1889 ; *br.* T. Brown & Son, *s.* Ajax 8,
d. M 851 by Paragon 268,
2nd d. by No. 2, 210.

M 202*a*, 202, lambed 1889 ; *br.* T. Brown & Son, *s.* Slasher 339,
d. Pedley's 6.

M 205*a*, 205, lambed 1889 ; *br.* T. Brown & Son, *s.* Leonard 173,
d. Paragon 268.

M 206*a*, 206, lambed 1889 ; *br.* T. Brown & Son, *s.* Ajax 8,
d. M 1045 by Slasher 339,
2nd d. by Sultan 352.

M 207*a*, 207, lambed 1889 ; *br.* T. Brown & Son, *s.* Ajax 8,
d. M 852 by Paragon 268,
2nd d. by No. 2, 210.

M 210a, 210, lambed 1889 ; *br.* T. Brown & Son, *s.* Slasher 339,
       d. M 897 by Sturdy 350,
       2nd d. by Broadfield 54.

M 215a, 215, lambed 1889 ; *br.* T. Brown & Son, *s.* Leonard 173,
       d. Lane's 2.

M 216a, 216, lambed 1889 ; *br.* T. Brown & Son, *s.* Ajax 8,
       d. M 1166 by Spot 348,
       2nd d. by Broadback 53.

M 222a, 222, lambed 1889 ; *br.* T. Brown & Son, *s.* Fyfield 122,
       d. M 614 by Beacon 34,
       2nd d. by Royal Kilburn 302.

M 235a, 235, lambed 1889 ; *br.* T. Brown & Son, *s.* Ajax 8,
       d. M 1173 by Shylock 315,
       2nd d. by Broadback 53.

M 236a, 236, lambed 1889 ; *br.* T. Brown & Son, *s.* Slasher, 339,
       d. M 1216 by Paragon 268,
       2nd d. M 585 by Example 102,
       3rd d. by Royal Kilburn 302.

M 239a, 239, lambed 1889 ; *br.* T. Brown & Son, *s.* Slasher 339,
       d. M 1219 by Paragon 268.
       2nd d. M 608 by Singleton 324,
       3rd d. by Royal Kilburn 302.

M 245a, 245, lambed 1889 ; *br.* T. Brown & Son, *s.* Oliver 264,
       d. M 598 by Example 102,
       2nd d. by Garne's Blackfoot 123.

R. S. 1, 1, aged ; *br.* R. Swanwick, *s.* No. 10, 231.

R. S. 2, 2, aged ; *br.* R. Swanwick, *s.* No. 10, 231.

R. S. 3, 3, aged ; *br.* R. Swanwick, *s.* No. 10, 231.

R. S. 4, 4, aged ; *br.* R. Swanwick, *s.* No. 10, 231.

R. S. 4, 4, aged ; *br.* R. Swanwick, *s.* No. 10, 231.

R. S. 5, 5, aged ; *br.* R. Swanwick, *s.* No. 10, 231.

R. S. 6, 6, aged ; *br.* R. Swanwick, *s.* No. 10, 231.

R. S. 7, 7, aged ; *br.* R. Swanwick, *s.* No. 10, 231.

R. S. 8, 8, aged ; *br.* R. Swanwick, *s.* No. 10, 231.

R. S. 9, 9, lambed 1889 ; *br.* R. Swanwick, *s.* No, 10, 231.

R. S. 10, 10, lambed 1889 ; *br.* R. Swanwick, *s.* No. 10, 231.

R. S. 11, 11, lambed 1889 ; *br.* R. Swanwick, *s.* No. 10, 231.

R. S. 12, 12, lambed 1889 ; *br.* R. Swanwick, *s* No. 10, 231.

R. S. 13, 13, lambed 1889 ; *br.* R. Swanwick, *s.* No. 10, 231.

R. S. 14, 14, lambed 1889 ; *br.* R. Swanwick, *s.* No. 10, 231.

R. S. 15, 15, lambed 1889 ; *br.* R. Swanwick, *s.* No. 10, 231.

R. S. 16, 16, lambed 1889 ; *br.* R. Swanwick, *s.* No. 10, 231.

R. S. 17, 17, lambed 1889 ; *br.* R. Swanwick, *s.* No. 10, 321.

R. S. 18, 18, lambed 1889 ; *br.* R. Swanwick, *s.* No. 10, 231.

R. S. 19, 19, lambed 1889 ; *br.* R. Swanwick, *s.* No. 10, 231.

R. S. 20, 20, lambed 1889 ; *br.* R. Swanwick, *s.* No. 10, 231.

R. S. 21, 21, lambed 1889 ; *br.* R. Swanwick, *s.* No. 10, 231.

R. S. 22, 22, lambed 1889 ; *br.* R. Swanwick, *s.* No. 10, 231.

R. S. 23, 23, lambed 1889 ; *br.* R. Swanwick, *s.* No. 10, 231.

R. S. 24, 24, lambed 1889 ; *br.* R. Swanwick, *s.* No. 10, 231.

R. S. 29, 29, lambed 1889 ; *br.* R. Swanwick, *s.* No. 39, 252.

R. S. 30, 30, lambed 1889 ; *br.* R. Swanwick, *s.* No. 39, 252.

R. S. 31, 31, lambed 1889 ; *br.* R. Swanwick, *s.* No. 39, 252.

R. S. 32, 32, lambed 1889 ; *br.* R. Swanwick, *s.* No. 39, 252.

R. S. 33, 33, lambed 1889 ; *br.* R. Swanwick, *s.* No. 39, 252.

R. S. 34, 34, lambed 1889 ; *br.* R. Swanwick, *s.* No. 39, 252.

R. S. 35, 35, lambed 1889 ; *br.* R. Swanwick, *s.* No. 39, 252.

R. S. 36, 36, lambed 1889 ; *br.* R. Swanwick, *s.* No. 39, 252.

R. S. 37, 37, lambed 1889 ; *br.* R. Swanwick, *s.* No. 39, 252.

R. S. 38, 38, lambed 1889 ; *br.* R. Swanwick, *s.* No. 39, 252.

R. S. 39, 39, lambed 1889 ; *br.* R. Swanwick, *s.* No. 39, 252.

R. S. 40, 40, lambed 1889 ; *br.* R. Swanwick, *s.* No. 39, 252.

R. S. 100, 100, lambed 1889 ; *br.* R. Swanwick, *s.* No. 14, 240.

R. S. 101, 101, lambed 1889 ; *br.* R. Swanwick, *s.* No 14, 240.

R. S. 102, 102, lambed 1889 ; *br.* R. Swanwick, *s.* No. 14, 240.

R. S. 103, 103, lambed 1889 ; *br.* R. Swanwick, *s.* No. 14, 240.

R. S. 104, 104, lambed 1889 ; *br.* R. Swanwick, *s.* No. 14, 240.

R. S. 105, 105, lambed 1889 ; *br.* R. Swanwick, *s.* No. 14, 240

R. S. 106, 106, lambed 1889 ; *br.* R. Swanwick, *s.* No. 14, 240.

R. S. 107, 107, lambed 1889 ; *br.* R. Swanwick, *s.* No. 14, 240.

R. S. 108, 108, lambed 1889 ; *br.* R. Swanwick, *s.* No. 14, 240.

R. S. 109, 109, lambed 1889 ; *br.* R. Swanwick, *s.* No. 14, 240.

R. S. 110, 110, lambed 1889 ; *br.* R. Swanwick, *s.* No. 14, 240.

R. S. 111, 111, lambed 1889 ; *br.* R. Swanwick, *s.* No. 14, 240.

R. S. 112, 112, lambed 1889 ; *br.* R. Swanwick, *s.* No. 14, 240.

R. S. 126, 126, aged ; *br.* R. Swanwick, *s.* Son of No. 9, 343.

R. S. 127, 127, aged ; *br.* R Swanwick, *s.* Son of No. 9, 343.

R. S. 128, 128, aged ; *br.* R. Swanwick, *s.* Son of No. 9, 343.

R. S. 129, 129, aged ; *br.* R. Swanwick, *s.* Son of No. 9, 343.

R. S. 130, 130, aged ; *br* R. Swanwick, *s.* Son of No. 9, 343.

R. S. 131, 131, aged ; *br.* R. Swanwick, *s.* Son of No. 9, 343.

R. S. 132, 132, aged ; *br.* R. Swanwick, *s.* Son of No. 9, 343.

R. S. 133, 133, aged ; *br.* R. Swanwick, *s.* Son of No. 9, 343.

R. S. 134, 134, aged ; *br.* R. Swanwick, *s.* Son of No. 9, 343.

R. S. 153, 153, aged ; *br.* R. Swanwick, *s.* Northampton First 202.

R. S. 154, 154, aged ; *br.* R. Swanwick, *s.* Northampton First 202.

R. S. 155, 155, aged ; *br.* R. Swanwick, *s.* Northampton First 202.

R. S. 156, 156, lambed 1889 ; *br* R. Swanwick,
                          s. Northampton First 202.

R. S. 157, 157, lambed 1890 ; *br.* R. Swanwick,
                          s. Northampton First 202.

R. S. 158, 158, aged ; *br.* R. Swanwick, *s.* Son of No. 9, 343.

R. S. 159, 159, aged ; *br.* R. Swanwick, *s.* Son of No. 9, 343.

R. S. 160, 160, aged ; *br.* R. Swanwick, *s.* Son of No. 9, 343.

R. S. 161, 161, aged ; *br.* R. Swanwick, *s.* Son of No. 9, 343.

R. S. 162, 162, aged ; *br.* R. Swanwick, *s.* Son of No. 9, 343.

R. S. 163, 163, aged ; *br.* R. Swanwick, *s.* Son of No. 9, 343.

R. S. 164, 164, aged ; *br.* R. Swanwick,
                          s. The Fifty Guinea Ram 365.

R. S. 165, 165, aged ; *br.* R. Swanwick,
                          s. The Fifty Guinea Ram 365.

R. S. 166, 166, aged ; *br.* R. Swanwick,
                          s. The Fifty Guinea Ram 365.

R. S. 167, 167, aged ; *br.* R. Swanwick, *s.* Royal Newcastle 303.

R. S. 168, 168, aged ; *br.* R. Swanwick, *s.* Royal Newcastle 303.

R. S. 172, 172, aged ; *br.* R. Swanwick, *s.* No. 9, 228.

R. S. 173, 173, aged ; *br.* R. Swanwick, *s.* No. 9, 228.

R. S. 174, 174, aged ; *br.* R. Swanwick, *s.* No. 9, 228.

R. S. 175, 175, aged ; *br.* R. Swanwick, *s.* No. 9, 228.

R. S. 176, 176, aged ; *br.* R. Swanwick, *s.* No. 9, 228.

R. S. 177, 177, aged ; *br.* R. Swanwick,
                          s. No. 9, 228, or Northampton First 202.

R. S. 180, 180, aged ; *br.* R. Swanwick, *s.* No. 40, 255.

INDEX OF
BREEDERS, OWNERS, & NOS. OF RAMS.

Breeders includes Owners where the Ram referred to has not changed hands.

	BREEDERS' NOS. OF RAMS.	OWNERS' NOS. OF RAMS.
AYLMER, HUGH	39, 88, 130, 140, 169, 176, 178, 257, 354, 355, 383, 384, 392, 393, 397	112, 244, 306
AYLMER, J. B.	112	
BAGNALL, G., & SON ...	41, 44, 155, 363	12, 36, 55, 197, 317, 318, 319
BARTON, CHARLES ...	20, 23, 24, 31, 32, 33, 36, 42, 43, 45, 47, 52, 64, 67, 79, 82, 85, 89, 90, 91, 92, 115, 116, 121, 122, 132, 134, 135, 148, 153, 161, 162, 180, 186, 191, 195, 196, 197, 198, 213, 217, 224, 228, 229, 235, 236, 237, 249, 259, 260, 262, 271, 280, 281, 293, 298, 307, 311, 322, 323, 326, 327, 328, 336, 338, 357, 358, 376, 396	11, 56, 230, 242, 245, 251, 258, 321
BEAK, GEORGE	... ...	213, 217, 224, 235, 249
BRAIN, EDWARD	29, 59, 192, 208, 214, 341	52, 149
BROWN, THOMAS	6, 28, 34, 51, 53, 61, 62, 66, 70, 71, 72, 75, 76, 77, 86, 87, 94, 96, 99, 101, 104, 109, 126, 128, 129, 136, 142, 143, 144, 146, 159, 163, 177, 185, 189, 190, 199, 212, 218, 222, 246, 264, 268, 278, 279, 283, 284, 292, 294, 297, 300, 302, 309, 313, 315, 324, 337, 339, 344, 345, 346, 347, 348, 350, 352, 353, 361, 367, 374, 385	3, 48, 54, 123, 127, 141, 164, 166, 181, 201, 219, 223, 225, 226, 234, 238, 256, 270, 329, 335, 370, 390

	BREEDERS' Nos. OF RAMS.		OWNERS' Nos. OF RAMS.
BROWN, T., & SON... ...	9, 10, 19, 60, 107, 110, 113, 138, 147, 174, 273, 277, 285		7, 8, 18, 85, 122, 173, 175, 265
BURRELL, Mrs.	... ...		381
BYNG, The Honble. Major L. ...	... ...		29, 59
CLARK, C. E.	179		46, 152, 184, 378
CLARK, HENRY	46, 118, 145, 152, 157, 184, 378, 379		79
CRADDOCK, F.	95, 360, 364		43, 369
CRADDOCK, JOSEPH ...	386		
ELWES, H. J.	69, 395		165, 194, 236, 275
FLETCHER, GEORGE ...	241		253
FOWLER, EDWARD ...	17, 258		
FREEMAN, GEORGE ...	291, 401		58, 187, 254, 314, 326, 402
GARNE, JOHN	111		153, 272, 387
GARNE, ROBERT ...	1, 3, 4, 5, 7, 8, 11, 12, 13, 16, 18, 37, 38, 49, 63, 65, 78, 81, 98, 100, 102, 103, 105, 106, 108, 120, 123, 124, 125, 127, 131, 139, 154, 181, 182, 183, 187, 193, 194, 203, 204, 209, 211, 215, 223, 225, 227, 230, 231, 232, 233, 234, 238, 239, 243, 254, 265, 267, 269, 270, 272, 274, 276, 282, 287, 288, 289, 296, 301, 314, 331, 340, 362, 369, 371, 372, 377, 379, 387, 390, 391, 394, 400, 402		30, 61, 88, 92, 104, 186, 189, 200, 221, 237, 247, 263, 316, 368
GILLETT, CHARLES ...	... ...		132, 137
GILLETT, J.	21		
GILLETT, THOS.	160, 370		

	Breeders' Nos. of Rams.				Owners' Nos. of Rams.
WAKEFIELD, J. P. ... {	149	...	...	...	49, 95, 108, 170, 341
WALKER, JAMES ...	329	...	...	...	330
WALKER, THOMAS ...	25, 83	...	...	...	233
WILLIAMS, O. F. ...	150, 381	...	...	...	35, 50
YEOMANS, J. H.... ...	114, 117, 119, 382			...	1, 2, 14, 15, 17, 20, 23, 57, 121, 151, 168, 171, 172, 193

MEMBERS OF
THE COTSWOLD SHEEP SOCIETY.

Those marked **C.** are Members of Council, **E.** Editing Committee, **F.** Finance Committee.

	Acock, Arthur ...	... Cold Aston, Cheltenham, Glos.
	Attwater, J. N. ...	... Sierford, Andoversford, Glos.
	Aylmer, Hugh ...	... West Dereham Abbey, Norfolk.
C.	Bagnall, T. O. ...	... Westwell Manor, Burford, Oxon.
C. E.	Barton Charles ...	... Fyfield, Lechlade, Glos.
	Beak, George ...	... Stamford Hall, Lechlade, Glos.
	Biddulph, Michael ...	Kemble, Cirencester, Glos.
C.	Brain, Edward ...	... Manor Farm, Upper Slaughter, Bourton - on - the - Water, Gloucestershire.
C. E. F.	Brown, Thomas *(President-Elect, Chairman of Editing Committee)*	Marham Hall, Downham Market, Norfolk
	Byng, The Hon. Major L.	Sherborne House, Northleach, Gloucestershire.

C. F.	CLARK, CHAS. ED. ...	The Down, Chalford, Stroud, Gloucestershire.
	CLARK, HENRY	Frampton Mansell, Chalford, Stroud, Gloucestershire.
	COOK, THOMAS	Taddington Broadway, Worcestershire.
	CRADDOCK, ERNEST ...	Eastleach, Lechlade, Glos.
	CRADDOCK, FREDERICK...	Eastington, Northleach, Glos.
	ELDON, The Right Hon. the Earl of	Stowell Park, Northleach, Glos.
C. E.	ELWES, H. J....	Colesborne Park, Andoversford, Gloucestershire.
	FLETCHER, WILLIAM H.	Shipton, Andoversford, Glos.
	FOWLER, EDWARD POPE	Aston Avening, Stroud, Glos.
	FREEMAN, GEORGE ...	Sherborne, Northleach, Glos.
	GARDNER, THOMAS ...	Clapton, Bourton-on-the-Water, Gloucestershire.
	GARNE, JOHN	Filkins, Lechlade, Glos.
C. E. F.	GARNE, ROBERT (President)	Aldsworth, Northleach, Gloucestershire.
	GARNE, WILLIAM... ...	Manor Farm, South Cerney, Cirencester, Glos.
	GARNE, WILLIAM T. ..	Aldsworth, Northleach, Glos.
	GILLETT, CHARLES ...	Lower Haddon, Bampton, Faringdon, Berks.
C.	GODWIN, J. J.	Troy Farm, Somerton, Banbury, Oxon.
	HANDY, AUBREY	Coln St. Dennis, Northleach, Gloucestershire.
	HANDY, EDWARD	Shipton, Andoversford, Glos.
	HATHAWAY, R.	Cotswold Farm, Duntisbourne Cirencester, Glos.

	HOULTON, H. T.	Taynton, Burford, Oxon.
	HOULTON, WILLIAM ..	Broadfield, Northleach, Glos.
	HUCKVALE, J. EVANS ..	Bruern Grange, Chipping Norton, Oxon.
C. E.	HULBERT, T. R.	North Cerney, Cirencester, Gloucestershire.
C. F.	JACOBS, ROBERT	Signett Hill, Burford, Oxon.
C. F.	PORTER, THOMAS *(Chairman of Finance Committee)*	Baunton. Cirencester, Glos.
	POWELL, WALTER... ...	Upton, Burford, Oxon.
	SMITH, W. J.	Hill House, Shilton, Bampton, Oxen.
C. E.	SWANWICK, RUSSELL ...	Royal Agricultural College Farm, Cirencester, Glos.
	TAYLOR, JAMES	Rendcomb Park, Cirencester, Gloucestershire.
	WAKEFIELD, JOHN P. ...	Great Barrington, Burford, Oxon.
	WALKER, THOMAS ...	Witney Street, Burford, Oxon.
	WILLIAMS, O. F.	The Weir End, Ross, Herefordshire.
C. E.	YEOMANS, JOHN H. ...	Stretton Court, Hereford.

PRINTED BY BAILY AND SON, CIRENCESTER.